COOPERATIVES, GRASSROOTS DEVELOPMENT, AND SOCIAL CHANGE

EDITED BY
MARCELA VÁSQUEZ-LEÓN,
BRIAN J. BURKE, AND
TIMOTHY J. FINAN

COOPERATIVES, GRASSROOTS DEVELOPMENT, AND SOCIAL CHANGE

Experiences from Rural Latin America

WITH TRANSLATIONS BY

MAISA C. TAHA

THE UNIVERSITY OF
ARIZONA PRESS
TUCSON

The University of Arizona Press
www.uapress.arizona.edu

Printed in the United States of America
22 21 20 19 18 17 6 5 4 3 2 1

ISBN-13: 978-0-8165-3474-6 (cloth)

Cover design by Nicole Hayward
Cover photo by Daniel Duncan

The chapter by Elisa Echagüe in the Paraguay section and the four chapters in the Colombia section were originally written in Spanish and were translated by Maisa C. Taha. The chapter by Joana Laura Marinho Nogueira and João Nicédio Alves Nogueira in the Brazil section was originally written in Portuguese and was translated by Timothy J. Finan.

Publication of this book is made possible in part by a subvention from the Bureau of Applied Research in Anthropology and the Center for Latin American Studies at the University of Arizona.

Library of Congress Cataloging-in-Publication Data are available from the Library of Congress.

∞ This paper meets the requirements of ANSI/NISO Z39.48-1992 (Permanence of Paper).

CONTENTS

FOREWORD

CHARLES C. COX

THIS PUBLICATION ADDRESSES ISSUES AND OPPORTUNITIES that present themselves to smallholder agricultural cooperatives in selected regions of Paraguay, Brazil, and Colombia. The context of the discussion is how these enterprises react to confront the challenges of operating successfully in a global economy that will reward winners and leave poor performers out of the marketplace. This process is characterized as management of change, an adaptation exercise all businesses and organizations must embrace to prosper in complicated and often rapidly changing circumstances.

In many regions of the world, cooperatives and other agribusinesses are flourishing because, among many positive factors, they provide products and services for which there is demand, they are sufficiently capitalized and have ready access to credit facilities, they receive substantial governmental support through favorable legal and regulatory frameworks that promote cooperative development instead of imposing state control on these entities, and there are technical and technological assistance providers locally available to these producer groups. Internally, successful cooperatives in prosperous regions can recruit professional managers and competent support staff from a robust human resource base, and their offices are equipped with the requisite technological tools. They possess a rich pool of strong candidates to compose informed and involved boards of directors, and they can finance member training to strengthen farmer participation in achieving the financial objectives of the enterprise.

The majority of cooperatives participating in the management of change analysis presented in this publication can count on very few of the positive factors cited above to help them flourish. They do have marketable products, and Brazil, Paraguay, and Colombia enjoy favorable legal and regulatory atmospheres for cooperative development. However, the cooperatives are located in remote areas of the Brazilian Amazon region and in some of the most poverty-stricken locations in eastern Paraguay, and, in the case of Colombia, some of the cooperatives are located in regions where there is illegal crop production. One might question why this USAID-funded worldwide Cooperative Development Program (CDP) activity (2004–2010) even targeted such locales for assistance and analysis.* The response is that USAID, the Bureau of Applied Research in Anthropology (BARA) at the University of Arizona, and the Agricultural Cooperative Development International and Volunteers in Overseas Cooperative Assistance (ACDI/VOCA) prioritize worldwide efforts to help small farmers increase family income and quality of life, hopefully reducing the exodus to cities of the rural poor—especially young people.

The selection of participating cooperatives in the assessments conducted by BARA, all limitations notwithstanding, centered on ACDI/VOCA's determination that indeed the cooperatives were producing marketable products and, crucially, that their management and boards of directors embraced the concept of adapting their thinking and actions to successfully confront demanding market conditions and challenges. It is important to note that the term "marketable products" used here connotes that the participating cooperatives are currently marketing one or more products. On the positive side, they are marketing internally, and some of them have found markets abroad, including in the United States, Japan, Western Europe, Russia, Uruguay, and Argentina. Issues confronting them on the exporting front include increasing production volumes, improving quality controls, and securing new markets.

*All of the cooperative experiences presented in the Paraguay and Brazil sections of this volume were assessed as part of the USAID-funded worldwide CDP activity (2004–2010). Of the cooperative experiences presented in the Colombia section, only that of Cuéllar and Ramírez was part of the CDP project. The chapters by Dávila and López are based on research at the Universidad de los Andes and the chapter by Ruiz is based on independent research.

The more complex side of the equation is the management of change. While the leadership and management of the cooperatives selected were interested in participating in the BARA assessments of their history, traditions, capabilities, capacities, and inclinations to survive and hopefully grow stronger, those statements of interest thereby engaged the leaders and managers in a process that could take them down a challenging path. Management measures deemed necessary, in some cases, might focus on management shortcomings, but it is recognized that many of these cooperatives are located in remote regions that don't have a large presence of trained managers and other professionals. Furthermore, the smaller cooperatives don't have the financial resources to employ well-qualified professional staff or adequately equip their offices. A reconfiguration of a dormant board of directors might be in order, but such efforts can be complicated; board members are recognized local opinion leaders, and they might not want to surrender their position of status in the community. A merger of two small, economically unviable cooperatives producing the same product in the same county might result in a stronger, competitive enterprise with the ability to attract more qualified staff, access credit, gain some economies of scale, and expand markets. The problem: the merger would result in the unwelcome redundancy of one manager and one board chairman and perhaps several board members.

The BARA interviewees at the village level have listed the plethora of roadblocks in their path, several of which are mentioned above. However, these managers and leaders have stated their inclination to adapt to changing circumstances despite the numerous limitations. The findings of BARA published here—in conjunction with those of ACDI/VOCA in cooperative organizational/managerial strengthening and value chain enhancement in the production, postharvest handling, processing, and marketing with these cooperatives—will be valuable in instructing ACDI/VOCA's approaches in the field during future CDP cycles. The program will continue to offer senior-level technical assistance to the cooperatives that are taking strides to modernize and compete successfully within the means available to them.

BARA's efforts in delving into the minds of cooperative managers, leaders, and farmer-members in a research exercise was probably the first time such assessments were incorporated in an ongoing technical assistance program delivering organizational and value-chain strengthening with the same cooperatives. The BARA component was unique, as it led the interviewees to internalize and process their reason for being and then look beyond such issues as whether the terrible roads in their regions would be passable when their bananas had to be shipped to Buenos Aires within ten days. The cooperative members were proud to recount their histories, downturns, and near collapses, as well as the resurrections and restructurings that brought them through tough times. These remembrances fortified them and seemed to have reinforced their determination to fight on and succeed. They managed to navigate through severe changes and setbacks over the decades. Why should they stop now?

ACKNOWLEDGMENTS

WE WOULD LIKE TO EXPRESS our sincere appreciation to all the generous and supportive people of the communities in Paraguay, Brazil, and Colombia who made this book possible. We are grateful to those cooperative leaders, members, and their families who participated in our studies for patiently answering our questions and enriching our experience.

We are also grateful for the assistance of our counterparts at ACDI/VOCA–Paraguay, Brazil, and Colombia. Special thanks to Amelia Moro and Nene Barrios in Paraguay, Dr. Celso Claro de Oliveria and Daniel Silva in Brazil, and Martha E. Gómez in Colombia.

We also thank the research assistants from the Bureau of Applied Research in Anthropology at the University of Arizona who contributed to our understanding of the cooperatives studied but are not authors of individual chapters. These include Paola Canova, Olga Lucia Cuéllar, Fernando Mayorga, Krisna Ruette, Anita Carrasco, and Lucero Vásquez-Radonic. We also appreciate the assistance of the students from the Department of Ecología Humana, Facultad de Ciencias Agrarias, Universidad Nacional de Asunción in Paraguay who accompanied us in the field. Thanks also go to María Rodríguez for helping us manage the project, to Cari Tusing for her editorial comments, and to Daniel Duncan for lending his visual expertise and photos.

A final special thanks to Maisa C. Taha for her diligence and tremendous patience in translating, editing, and providing critical insight that considerably improved several chapters in this book.

COOPERATIVES, GRASSROOTS DEVELOPMENT, AND SOCIAL CHANGE

South America: Case Study Cooperatives in Paraguay, Colombia, and Brazil

INTRODUCTION

SMALLHOLDER COOPERATIVISM AS A DEVELOPMENT STRATEGY IN LATIN AMERICA

MARCELA VÁSQUEZ-LEÓN, TIMOTHY J. FINAN, AND BRIAN J. BURKE

THROUGHOUT LATIN AMERICA, smallholder agriculture remains a significant part of the regional agricultural economy. There are an estimated 15 million family farms in Latin America occupying about 400 million hectares. These farms have been characterized into three types: (a) an estimated 10 million subsistence farms that provide neither adequate incomes nor food security and are part of diversified livelihood strategies of mostly poor households; (b) around 4 million farms that have a broader asset base and are integrated into local markets but face significant environmental and institutional constraints; and (c) about a million smallholder farms with adequate assets and supporting institutional contexts (Berdegué and Fuentealba 2011). While rural populations have declined in the face of a rural–urban exodus, in an especially marked way in South America (United Nations 2014), poverty remains entrenched among rural households within the family farm subsector. Moreover, while in many countries rural households have benefited from social transfer programs and increased education and health access, public investment in smallholder agriculture through credit, input, and marketing incentives has decreased.

With the rise of market-based strategies for development, Latin American agriculture has seen its export sector increase in both traditional export crops (coffee, cocoa, sugar, etc.) and nontraditional commodities (fruits and vegetables, flowers, soybeans, and meat). The farm systems that participate in export-based markets tend to be large-scale agribusiness firms and not smallholder farms, although these latter play an exceedingly important role in the supply of domestic markets. These recent trends of agricultural change frame the basic challenge of Latin American agricultural development. Since smallholder agriculture seems unprepared to compete in an ever-growing export market dominated by agribusiness firms, it remains a sink of rural poverty with few escape options under current models of growth. At the same time, urban, nonagricultural sectors have not absorbed the inflow of rural labor in ways that would permit a smooth transfer into the urban middle class (Ocampo and Martín 2003). While indicators of rural development (e.g., food production and commodity export) appear promising on average, these national averages mask deep regional and local disparities wherein the concentrations of rural poverty are stubbornly resistant.

Participation in expanding markets is thought to provide a pathway to agricultural growth and to increases in rural incomes. It is fair to assert that many, if not most, smallholder producers are well integrated in local and regional markets and are attracted to global opportunities. Nonetheless, smallholders, by definition, do not achieve economies of scale and are highly vulnerable in the competitive environment dominated by large and corporate firms, many of them multinationals. In fact, entry into the global export market can present unprecedented levels of risk for smallholders unfamiliar with global price variability, highly demanding quality standards, rigid deadlines, government regulations, and so forth. As a further constraint, such farmers seldom can generate the capital and credit needed to compete with larger firms. Thus, as individual firms, smallholders face a decided disadvantage in global markets.

The thesis of this book is that a possible avenue of inclusion lies in cooperative organization. The development community widely shares the conviction that associativism and collective action can create pathways of access to global markets for smallholder producers, especially to markets for niche and specialty items (e.g., fair trade coffee and organic sugar). Using the strategy of cooperation, smallholders are able to achieve economies of scale, attract wholesale credit sources, and, just as important, create market visibility through branding and product differentiation. Thus, we focus here on the development role of smallholder agricultural cooperatives, asking in what ways and under what conditions do cooperatives provide smallholders access to a more productive and remunerative livelihood as well as a viable forum through which they can engage politically in the transformation of their own communities.

The tradition of formal cooperativism reaches back over 160 years, and the original principles of organization represent the core values of what in development is now called "good governance." Cooperatives not only confer economic advantages (e.g., scale economies) but also provide a model of democratic and transparent decision making, of participation blind to social class, gender, or race, and of holistic well-being (i.e., improved education, health, access to information, professional identity, etc.). These principles in the Latin American context are particularly important, since the poor in Latin America are not just economically poor, but voiceless, marginalized, and socially excluded. To examine smallholder cooperatives in Latin America, then, is to test whether the model of cooperative organization is a viable alternative to large-scale economic growth, to test the social inclusiveness of cooperatives, and to evaluate the comprehensive social development of cooperative members and their local communities.

In this sense, we target a more specific group. It has already been established that in the more economically advantaged areas of Latin America, such as the wine country of Chile or the southern states of Brazil, cooperatives extend an obvious advantage, and they are widely present and active. Here cooperatives are gateways to world trade, well-recognized brand names, and vertically integrated value chains. In the smallholder and more impoverished regions, however, cooperativism registers more moderate successes, at times even undesirable impacts that exacerbate the inequities woven into the social cloth. Our focus, then, is on the broader development potential of cooperativism as a form of social organization, as a sustainable tool for enhancing the well-being of the underresourced and the socially, as well as politically, excluded so as to counteract the deeply rooted stratification that characterizes Latin American rural society.

In the chapters presented here, we describe a set of contemporary cooperatives in poor smallholder regions of Paraguay, Brazil, and Colombia, examining their strategies for adapting to a rapidly changing world. Smallholder rural societies in Latin America are often immersed in highly complex economic and social relationships that are transnational and global (Kearney 1996). At the same time, they continue to represent values that center on place and community, where the goal is to achieve viable and stable livelihoods in the community and on the land (Netting 1993; Fals Borda 2002). In each case study, we examine the history of the individual cooperative and identify the processes that have determined its trajectory of change. We address the central question of how cooperative organization in a highly stratified society is able to negotiate the local context of inequality and the broader context of international markets and global competition. In other words, do cooperatives have the potential to significantly alter the unfavorable balance of power and "level the playing field" for so many small producers in Latin America? As these chapters show, in eastern Paraguay, northern Brazil, and rural Colombia, inequality has deep historical roots and manifests itself today in smallholders' limited farm size, in their occupation of marginal land and insufficient access to credit, and in a general lack of public investment in physical and social infrastructure.

We approach cooperatives as dynamic community-based organizations that exist within wider social, political, and economic contexts. As collective organizations, their theoretical advantage is derived from improved access to a variety of assets including economic capital (credit, land, and financial investment), social capital (networks that improve collective bargaining and marketing potential), and human capital (investment in member education, health, capacity building, and professionalization). In practice, however, the success of individual cooperatives is highly dependent on a number of factors, including the competency of internal management and the vicissitudes of external market and policy forces. In order to sustain themselves, cooperatives must have the ability to adapt to these highly complex institutional environments while, at the same time, they respond to the socioeconomic needs of an often marginalized and resource-scarce membership.

Our interest in assembling this book stems from a multiyear research project that has examined agricultural, fishing, and extractivist cooperatives representing a wide variety of structural characteristics such as size, function, commodity, commercial and financial stability, and ethnic composition. Within this diversity, we employ a common analytical framework that integrates five key contemporary issues:

1. The essential trade-off between financial success and social investment in the membership: When the cooperative boasts a well-educated, healthy, and financially stable membership, the role of the cooperative is fundamentally the commercial one of increasing market access. In contrast, our case-study cooperatives for the most part have a membership that is resource-scarce, undereducated, and dominated by local power elites. In this context, the cooperative plays a much more complex role, involving not only market access but often such things as education, technology transfer, and health services. How does a cooperative model thus create economic success while maintaining social solidarity and the broader goals of empowerment and inclusion?

2. The critical role of the state (and state policy) in the success of cooperatives: Cooperatives are state-sanctioned organizations subject to a complex set of policies that either directly or indirectly affect cooperative decision making, management, investment, and outcomes. Cooperatives are also well positioned to represent the interests of smallholders at the national policy level. The case studies presented here explore how public policies have shaped the historical trajectory of each cooperative by introducing both opportunities and constraints. They also examine the potential of smallholder cooperatives to take a proactive political role in pursuing policies in their favor.
3. The interaction of local cultural patterns and change strategies: The stark reality of social stratification is ingrained in cultural values and traditions that work against the democratic principles of cooperativism. Such values in Latin America are particularly embodied in patronage and clientelism. Cooperatives may be dominated by local elites, and more impoverished members may be reluctant to participate actively. Our analysis focuses on how such traditional values shape the management practices and impacts of cooperatives.
4. The effects of international market requirements on cooperative success: Integration into the international marketplace demands a "new" way of doing business. Particularly in the case of quality standards, certification, and product consistency, cooperatives must train their membership, invest in facilities, and gain access to sophisticated information systems in order to achieve and maintain market share. How do these market requirements advance or constrain the economic and social achievements of cooperatives?
5. Management and local governance: Fundamentally, a cooperative must nurture a sense of cooperative ownership among members and create a sense of collective identity and empowerment. At the same time, the cooperative management must have knowledge of good business practices and promote an entrepreneurial culture. Providing for the collective and acting in solidarity requires the ability to balance external advice with local capacity building, as well as the ability to apply internal controls, standards, pressures, and sanctions to act efficiently and effectively. Our case studies examine the challenges presented by this delicate balance between outward linkage and internal governance.

The chapters presented here cover a range of perspectives, including those of practitioners working with cooperatives, cooperative leaders themselves, and academic researchers. While each case is unique and illustrative in its own right, we intend to show that the cooperatives are adapting to a new uncertain environment, one which is part of the global movement seeking to integrate historically marginalized smallholders into regional, national, and global markets, networks, and power structures sometimes located far beyond their local communities. Whether these linkages can assure sustainable development or even guarantee participation and governance is not yet known. However, in the current Latin American political environment, where there has been an unprecedented increase in the influence of civil society in public decision making (Stahler-Sholk, Vanden, and Krueker 2008), smallholder cooperatives, as state-sanctioned institutions, can become key forms of association for strengthening Latin America's struggle for democratic forms of participation and social equity. As several scholars point out, this is particularly relevant as neither the state nor the private capitalist sector have been capable of making the necessary transformations to promote a more equitable rural society in Latin America (Davila et al 2005; Gonzales and Phillips 2013).

AGRICULTURAL COOPERATIVES IN LATIN AMERICA: THE CONTEXT

To understand the agricultural cooperatives presented in the different case studies, it is important to place them in the larger historical context of the development of cooperatives and cooperativism in Latin America. Contemporary production cooperatives have their roots in European and Asian systems of cooperation, starting with the waves of European and Japanese immigrants that arrived in the region in the nineteenth and twentieth centuries. Cooperatives were often created by immigrant communities and were instrumental in allowing them to re-create their own forms of social and economic organization while remaining detached from local populations and systems of government. Prime examples include the agricultural cooperatives in western Paraguay founded in the 1930s by Mennonites from northern Europe as well as the Japanese cooperatives in southern Brazil formed even earlier. Similar histories apply to the immigrant colonies of Italians and Germans in Brazil in the nineteenth century, whose contributions to the development of the agricultural economy of southern Brazil are widely acknowledged.

The European cooperative model was expanded across Latin America after World War II. Early initiatives followed a top-down development model, and concerns over inequality, widespread poverty, and social unrest throughout the region resulted in the promotion of production cooperatives in more marginal rural areas. Emphasis was placed on the need to create economies of scale by pooling resources and labor, expanding credit opportunities, and more equitably distributing benefits (McGrath 1978). Discussions quickly became politicized, however, and in many cases, rather than serving the needs of the poor, cooperatives became political organizations co-opted by national elites. As Bennett (1983, 6) notes, Western European cooperativism was "transformed into quite different forms" in Latin America. This was especially the case in marginalized regions, where democratic control was undermined by authoritarian and patron-client social structures in which family loyalties, class interests, or orders from local political bosses were given priority over the development of democratic governance. As Fals Borda (1971, 12) concluded from his study of cooperatives in Colombia, Venezuela, and Ecuador, "the cooperative movement was imposed from above as a paternalistic and authoritarian act; it was not the result of popular conviction based on democratic participation or popular enlightenment."

In socialist countries, cooperatives were seen as political organizations capable of radically changing the status quo by empowering the poorest segments of society. State-controlled interventions imposed collective production models aimed at increasing agricultural productivity, such as in Cuba, where state farms became the predominant organizational form. The expansion of educational opportunities and of the military and construction sectors after the 1959 Cuban Revolution led to a significant shortage of agricultural labor, resulting in the development of highly mechanized, state-controlled agriculture. Some of the major problems associated with state farms, as noted by Deere, Pérez, and Gonzalez (1993, 198), had to do with "low labor productivity and inefficiencies associated with an excessive degree of centralization and planning," as well as with the lack of local participation in decision making. In countries such as Mexico, with a strong populist government, state intervention in the collective structures of production reinforced hierarchical differences in wealth and skill. The embezzlement of funds by cooperative directors or the use of cooperatives as centers of political control by national parties were common occurrences, as in the case of Mexican fishing cooperatives and agricultural *ejidos* during the 1970s and '80s under the Institutional Revolutionary Party (Vásquez-León 1998).

Throughout the 1970s, the cooperative movement in general was criticized for its inability to deliver on promises of cooperation, development, democracy, participation, equity, solidarity, and social transformation. Gabriel Gagnon (1976), who studied cooperatives throughout Latin America and Africa, emphasized that cooperatives in capitalist societies abandoned their role as social movements to become better integrated into the dominant capitalist system while cooperatives in socialist countries became little more than vehicles for the transmission of state policy. Despite serious criticisms, however, scholars still valued cooperatives as organizations with the potential to challenge the status quo on behalf of the "legitimate interests" of the peasantry as a social class (Fals Borda 1971, 146–47).

Parallel to top-down approaches were grassroots small-scale industry, artisan, credit, savings, and housing cooperatives and associations that developed independently from the state (ILO 2001; Nash et al. 1976). Some of these grassroots organizations were influenced by the Catholic Church's liberation theology, which offered critical interpretations of the complex social realities surrounding rural communities. Ideas about social justice became operationalized through the participatory action and mutual support structure of the local ecclesiastical communities (Levine 1988). Grassroots organizations that developed under strong military dictatorships, such as the Christian Agrarian Leagues in Paraguay, were accused of promoting communism and were quickly silenced and violently repressed (Lewis 1984). Others, however, provided an important opportunity for impoverished producers and, despite attempts at repression, particularly in Brazil, remain viable today.

In the mid-1980s, the structural adjustment policies deployed by most Latin American governments began to have significant impacts on state farms and cooperatives. States promoted free trade and the privatization of productive and financial sectors, and significantly reduced economic protections once provided by subsidies, market protections, and export and tax privileges. In countries such as Mexico, large cooperative federations lacked the flexibility to adjust to new pressures and collapsed once external government assistance was withdrawn. Cooperatives were also dramatically affected by increases in transportation costs and input prices and the loss of basic social services (Kroeker 1996; Vásquez-León and Liverman 2004). In Cuba, the collapse of the socialist trading block led to the decentralization and dissolution of the state-farm sector. Collective organization, however, continued as Cuba's primary structure when state farms were converted to cooperatives. Even though state agencies continue to be the main buyers of outputs and suppliers of inputs, limiting cooperatives' management authority, there is still much greater member participation and on-farm decision making than in the past (Royce 2004). In Nicaragua, which adopted the cooperative model after the 1979 Sandinista revolution, cooperatives have continued to be important organizations despite—and to some extent because of—the decline in state support after the end of the Sandinista government in 1989. Ruben and Lerman (2005) attribute the continuation of the cooperative sector to the importance of social capital as a risk-sharing mechanism that ensures access to a variety of services no longer provided by the state.

For the more independent grassroots small-scale cooperatives and associations throughout Latin America, the withdrawal of state intervention was not as devastating (ILO 2001). Nonetheless, their ability to adjust to and transform their socioeconomic landscapes continues to be limited by more pervasive issues related to mass poverty, armed conflicts, political and economic instability, unemployment and underemployment, and environmental degradation. Concerns about these issues and the growing contradictions between the neoliberal globalization model and the claim to social rights (Eckstein and Wickham-Crowley 2003) are being addressed in the literature through innovative concepts of the *economia solidária* (social or solidarity economy),* which De Sousa Santos (2006) calls an alternative, counterhegemonic form of globalization, and of alternative arrangements for trade and exchange. For example, fair trade and various agricultural certification schemes constitute a growing movement that sees environmentally and socially responsible trade as a way in which small producers may successfully compete in international markets (Page-Reeves, 1998; Hernández Castillo and Nigh 1998; Grimes and Milgram, 2000). At the center of these efforts are collective organizations (i.e., cooperatives, producer associations, responsible consumer groups, ethical banks) perceived as the seeds of a fairer, more democratic, and more sustainable economy through which global capitalism can be transformed to address concerns for social justice and

*The idea of the *economia solidária* is that economic efficiency is measured not entirely by profitability but also in terms of quality of life (Laville and García Jané 2009, 1). Singer (2006, 3) defines solidarity economy as one in which those who work are also the owners of a particular enterprise, and there are no owners who are not also workers. In addition, all members have an equal voice in the decision-making process, property is equally divided, and the enterprise has broader goals of empowerment and inclusion. However, legal definitions vary across Latin America.

environmental sustainability (Healy 2001; Mutersbaugh et al. 2005). These viable alternative models of development foreground a renewed emphasis on the importance of community as well as efforts to transform community development through civic engagement and collective action (Marshall 2010; Gonzales and Phillips 2013).

It is the intrinsic character of cooperatives that they generate and must navigate numerous contradictions. In contrast to other social movements, such as the Zapatista autonomous communities in Chiapas, Mexico (Mora, 2008), or Brazil's Landless Workers' Movement (Vanden, 2008), which clearly depart from the existing legal and institutional order and often seek to destabilize or replace governments, cooperatives seek to work within the system that they want to change. As economic, social, and political organizations, they seek to bridge the local and the global and must be able to adjust to external policies while promoting self-reliance, democratic governance, sustainable generation of income, and deep commitment to community as defined by members (MacPherson, 2002). Even the notion of market-based social change is, as Murray and Raynolds (2007, 9–10) point out, contradictory, in that cooperatives work both within markets, seeking more egalitarian trade relations between northern consumers and southern producers, and against markets, looking to transform conventional trade practices that perpetuate inequalities. Mooney (2004) argues that these contradictions must be perceived as positive and functional characteristics that facilitate the development of dynamic, flexible, and responsive organizations.

In effect, two models of cooperativism in Latin America have emerged, and they form the backdrop of the messages contained in this volume. From a more politically conservative perspective, the role of the cooperative is restricted to the expansion of economic opportunity; once achieved, the social ameliorations (education, health, equity, and participation) will follow. The case studies in this volume, however, assert a more progressive analysis of this role: that cooperatives are seldom able to achieve economic goals without simultaneous improvements in education, health, social learning, and empowerment. These are interwoven and inextricable components of the dynamic that determines the ultimate success of cooperatives as development organizations for the smallholder communities. It is hoped that the reader, after reviewing the case studies presented here, will agree with our conclusion that the vast theoretical potential of cooperative action can only be realized in the practice of effective participation, shared learning, and solidarity.

BOOK ORGANIZATION AND A NOTE ON METHODS

This volume documents the stories of smallholder cooperatives in three Latin American countries—Paraguay, Brazil, and Colombia. Read as a whole, these stories convey the immense opportunities and challenges faced by small-producer cooperatives. The cooperatives profiled here were selected because they represent the unfiltered reality of cooperative development in a highly stratified society, because they reveal important "lessons learned," and because they have faced struggles adapting to a changing social, economic, and natural environment. Each major section is organized around a country, and for each country an introductory chapter is presented that provides a national context. Each section also contains contributions by practitioners working with cooperatives (Cox, Echagüe, Marinho Nogueira and Alves Nogueira, Cuéllar Gómez and Ramirez Anaya, and Ruiz Marín), academic researchers (Vásquez-León, Finan, Burke, Rentería-Valencia, Piekielek, Barros, Ladrón de Guevara, and López Cerón), and, in one case, a cooperative leader (González Aguilera).

In the Paraguay section, after a brief introduction, four case-study cooperatives are included. The case study presented by Burke details the story of a long-standing credit and agricultural cooperative, focusing on the lessons that can be learned from the cooperative's decline after almost three decades serving as a dynamic social and political force at the local and regional levels. The second case study, by Vásquez-Léon and González Aguilera, explores a sugar cooperative that, in a relatively short period of time, managed to obtain organic and fair trade certifications to enter global markets, break the abusive monopoly of the local sugar mill, and achieve greater control over the value chain. Rentería-Valencia's chapter describes the struggle of a banana cooperative seeking to stabilize production by entering Mercado Común del Sur (Southern Common Market—MERCOSUR) markets and addressing the larger structural problems and internal shortcomings that impact product quality. The final case study refers to a credit and agricultural cooperative that is trying to diversify agricultural production and target a variety of markets in order to address concerns about rural outmigration. This section ends with a concluding chapter by a Paraguayan cooperative practitioner, Echagüe, who provides a thoughtful reflection on the challenges currently facing smallholder producer cooperatives.

In Brazil, the contextual description is followed by four case studies. Piekielek and Finan tell the story of a highly diversified Japanese-Brazilian cooperative in the Amazon rain forest. By paying close attention to the cooperative's history, they highlight the achievements of the cooperative after it confronted multiple significant obstacles. Barros presents the case of an aquaculture cooperative raising tilapia for a regional market, an initiative started through state support for displaced sharecroppers. In the next chapter, Finan details a success story of cooperation in action by focusing on a small pineapple and *cupuaçu* cooperative located deep in the Amazonian region. In the final chapter of the Brazil section, Burke analyzes an indigenous cooperative processing and marketing Brazil nut products under an international agreement. He highlights considerations of power and history in analyzing the role of cooperativism as a form of collective organization that might transform local political economies.

The case studies presented in the Paraguay and Brazil sections are the result of multiyear research projects (2004 to 2009) conducted by a team of anthropologists from the Bureau of Applied Research in Anthropology (BARA) at the University of Arizona in collaboration with Agricultural Cooperative Development International and Volunteers in Overseas Cooperative Assistance (ACDI/VOCA), a Washington-based private nonprofit organization that promotes economic development by offering a wide range of technical assistance to cooperatives in developing countries. The strategy used to collect data was based on standard ethnographic methods and a team-based approach (see Vásquez-León, Burke, and Rádonic 2009). Throughout the project, the team collected primary data during several seasons of summer fieldwork. Initially, open-ended interviews were conducted with officials in public agencies, cooperative managers, and support institutions. Snowball sampling techniques were used to develop a research sample representing key economic and public service sectors. The data gathered at an institutional level allowed us to create a comprehensive framework that traced the institutional linkages between cooperatives, public institutions, and other economic agents.

To conduct research at the community level, we surveyed a statistically valid sample of at least 30 cooperative member households in each community. Household surveys focused on livelihood diversification strategies, views on cooperative performance, and impacts of the cooperative on household decision making. Qualitative methods (participant observation, semistructured key informant interviews, focus group interviews, and oral histories) were used to capture the range of perceptions about cooperative history, organizational structure, performance, change strategies, impacts, articulation with public and private sectors, and market integration. An ethnographic method allowed for a comprehensive approach that was comparative (across countries and different types of cooperatives), multilevel (local, regional, and global), and sensitive to historical dimensions.

In the Colombia section, the contextual chapter is followed by Cuéllar Gómez and Ramirez Anaya's analysis of the coffee cooperative sector and the impact of specialty coffees on the transformation of smallholder cooperatives. In the second chapter of this section, Ruiz Marín describes how a small and volatile cooperative became a significant economic force capable of transforming the national milk market. In the final chapter, López Cerón assesses savings and loan cooperatives established in the region of southern Santander with the assistance of the Catholic Church. The case studies presented in the Colombia section are also based on long-term research but follow a different methodological approach, one based more on economic analysis and long-term practitioners' reflections, as specified in each chapter. Throughout the book, the contributions of practitioners are the result of their own observations and reflections after many years of professional engagement with smallholder cooperatives in their respective countries.

In the conclusion, Vásquez-León and Burke synthesize the experiences across these three countries and identify the broader insights and lessons that can be scaled up and widely applied. These insights touch on several critical issues, including the need to translate economic gains into transformative social investments, the role of the state in instituting supportive policies and promoting an environment of solidarity, the importance of tailoring change strategies to local cultural patterns that must often also be challenged, the need for flexible strategies to ensure successful participation in international markets, and the importance of achieving effective leadership with strong participatory governance. They end by anticipating the challenges and opportunities that Latin American rural cooperatives will confront over the next few decades.

COOPERATIVE EXPERIENCES IN PARAGUAY

THE PARAGUAYAN CONTEXT OF SOCIAL INEQUALITY

MARCELA VÁSQUEZ-LEÓN

PARAGUAY IS THE MOST AGRARIAN of Latin American countries. Approximately 40 percent of its 6.5 million residents live in rural areas. Of these, 60 percent are smallholders who cultivate between 1 and 10 hectares of land, often practicing swidden agriculture and using animal traction to till the soil. Even though they occupy only 4 percent of the total agricultural land, these nearly 230,000 small farms generate over one third of the total value of national agricultural production and produce around 75 percent of the country's staple foods, making a significant contribution to food security (Sorrenson et al. 1998; IICA 2004; GIZ 2016).

Paraguay also stands out for its low socioeconomic indicators, ranking among the poorest countries in Latin America and Africa. In 2006, maternal and infant mortality rates stood at 186 per 1,000 live births compared to the regional average of 19.8 per 1,000 (World Bank 2006). Even though social spending increased during the 1990s, it amounts to only 6 percent of the country's GDP, compared to twice that for Latin America as a whole. In addition, Paraguay has the highest population growth rate in the region (2.5 percent), with 97 percent of the population concentrated in the country's eastern region (DGCE 2002). This region is composed of fertile plains with extensive valleys and lowlands where 98 percent of the farms are located, mostly smallholders who only occupy 4 percent of the total agricultural land (IICA 2004). The western region, a vast plain with large arid expanses known as El Chaco, accounts for 63 percent of Paraguay's total surface area and contains only 3 percent of the population (DGCE 2002). The region has the largest cattle ranches in the country, extending up to 50,000 hectares.

In order to understand the current situation of the four Paraguayan smallholder cooperatives presented here, as well as their members and the rural communities they serve, it is important to first provide a brief historical background of key agrarian policies that impacted the development of cooperatives and smallholder producers in Paraguay. Understanding this history is important to explain Paraguay's adverse and unequal socioeconomic indicators today, and to better appreciate the internal and external structural challenges that Paraguayan cooperatives have had to confront as they strive to become agents of socioeconomic transformation and local development. In the final chapter of this section, Echagüe places these cooperatives within the larger institutional context of cooperativism in Paraguay and provides a thoughtful reflection on the challenges facing smallholder producers.

A HISTORICAL PERSPECTIVE ON COOPERATIVES AND AGRARIAN DEVELOPMENT IN PARAGUAY

Historians date the first cooperative systems in Paraguay to the Jesuit *reducciones* of the seventeenth century, when work collectives were organized in indigenous communities to undertake agrarian activities. This type of collective work, known as the *minga*, dominated rural life for more than 100 years, until the expulsion of the Jesuits in 1767. In 1927, the Paraguayan state granted lands to Russian Mennonite immigrants in El Chaco. In exchange for colonizing this inhospitable frontier, Mennonites were left to organize and establish themselves independently from the rest of Paraguay. They created three agricultural cooperatives, which, up to the present, are well-known for their agricultural production and processing of dairy products and beef (Stoesz and Stackley 2000). Local indigenous populations were used by the Mennonites as a ready source of labor, and eventually, as Mennonite-Indian relations became more complex, the Mennonites formed the Association of Indian-Mennonite Cooperative Services (Asociación de Servicios de Cooperación Indígena-Mennonita—ASCIM) to proselytize and assist indigenous peoples in the region.

The first agricultural cooperative based on Rochdale principles, which provide the basic cooperative guidelines of democratic cooperative membership, cooperative education, autonomy, and concern for community, was established in 1941, when the Dirección General de Cooperativismo (Cooperativism's Central Committee), under the Ministerio de Agricultura y Ganadería (Ministry of Agriculture and Livestock—MAG), was created and the first Cooperative Law was formulated. The Cooperativa Ideal del Paraguay Limitada, as it was named, was established in the central area of the eastern region by a group of Paraguayans and German, Brazilian, and Arab immigrants for the production of grapes and the elaboration of wines (Rodriguez Silvero et al. 2005).

During the 35-year military dictatorship of Alfredo Stroessner (1954–1989), the state endorsed cooperative organization, with strong government intervention, as part of a larger agrarian reform. During this time, core agrarian policies focused on agro-exports that benefited the state, intermediaries, and Paraguayan and foreign *latifundistas* at the expense of peasant farmers (Lernoux 1982, 20–23; Formento 2003, 63–64). One of the first policies implemented in 1954 was the sale of large extensions of agricultural lands to multinational agribusinesses, particularly Brazilian and North American firms. These lands were primarily used for soy production, extensive cattle ranching, and timber extraction. The Paraguayan state assured these large farmers access to subsidized credit and the road infrastructure that would take crops, markets, and labor away from traditional routes and toward Brazilian ports, connecting remote areas with major markets. About one million hectares, or 30 percent of Paraguay's arable land, had been transferred to Brazilian settlers and multinationals by the beginning of the 1980s (Lernoux 1982, 21). The Stroessner government openly supported the creation of large cooperatives among these producers in the eastern region for the production and marketing of livestock and soy. While such immigration stimulated an agro-export boom, these better-capitalized large-scale farmers found land prices in Paraguay to be relatively cheap. This pushed land prices up, making it very difficult for poor Paraguayans to buy land. During this period, Paraguay's virgin forests, in a form of South-South resource colonialism, "became the new frontier for the Brazilian economic miracle of the 1960s and 1970s" (Weisskoff 1992, 1536).

In 1964, the Instituto de Bienestar Rural (Institute of Rural Welfare), or IBR, was created to plan and implement a massive internal colonization project called the Settlement of the Northern Axis. The project was designed to open the agricultural frontier to peasant farmers, reduce pressure on the overpopulated and overused lands in the central departments of eastern Paraguay, and increase foreign exchange by integrating the peasant farmer into the export-oriented growth model envisioned by the state. Through the IBR, the Stroessner government promised land titles, credit, and technical assistance to individual households. Beneficiaries, as Formento (2003) recounts, received strips of land measuring 200 by 1,000 meters, each plot located in engineering style along a road, with little regard for the social use of space. This, she notes, "determined a structure that threatened the communal bases by disarticulating preexisting relations of economic cooperation and integration among peasants" (2003, 66). Small farmers were left on their own to clear and cultivate the land, depending on highly unfavorable relations to suppliers and intermediaries for credit.

The continued impoverishment of small farmers in the newly colonized lands was further intensified by the lack of state investment in basic infrastructure and services such as the provision of potable water, electricity, schools, and health facilities. Farmers supplemented their earnings with off-farm work and seasonal migration, but many were eventually forced to abandon or sell their land illegally to work as laborers on large farms (Nagel 1999). This land was in turn purchased by agricultural investors, who would buy consecutive plots for large-scale production. Thus, while the colonization policy was distributing state-owned land, it was also encouraging private colonization and land concentration. Lernoux explains the situation of many of these farmers at the end of the 1970s: "Few of the peasants . . . ever learn to read or write, see a doctor, or know the luxury of running water or electricity. Malnutrition causes nine-tenths of the deaths . . . These peasants explain their tragedy by quoting an old Guarani Indian saying: 'No one listens to the cry of the poor or the sound of a wooden bell.' Three-fifths of Paraguay's 2.6 million people live this way" (Lernoux 1982, 20).

Throughout the 1970s, in tandem with the internal colonization project, the Paraguayan MAG promoted the creation of savings and credit cooperatives with the assistance of the U.S. Agency for International Development (USAID) and the Credit Union National Association (CUNA), a federation of credit unions in the United States. They envisioned cooperatives contributing to neighborhood improvement in small rural towns. During this period, many of the agricultural cooperatives, including three of the four cooperatives in our study, were initiated by rural schoolteachers, who incorporated a small percentage of local farmers. With mixed success, some of these cooperatives had a positive social impact in that they were able to finance some basic services, such as sanitation projects and the paving of neighborhood streets, which the state neglected to provide. In this sense, cooperatives served the interests of the state by allowing the state to minimize social spending and avoid redistributive policies.

In 1971 the Federación de Cooperativas de Producción Limitada (Federation of Production Cooperatives—FECOPROD) was created to promote the creation of new cooperatives, and a year later, a new Cooperative Law was issued that expanded the definition of cooperatives to include service cooperatives. In 1979 the Central Nacional de Cooperativas, CREDICOOP, was established with the fusion of 44 credit and agricultural cooperatives. Its primary objective was, and continues to be, the provision of financial and technical assistance to its members. By 1980 there were 166 registered cooperatives, of which 98 were agricultural producer cooperatives. There were also 50 savings and credit, 10 consumer, and 8 service cooperatives (MAG-INCOOP 1997).

As part of the overall strategy to expand agro-exports, the state deployed a policy of full control over cooperative associations in rural areas, which prohibited small farmers from organizing in any form that "sought to counter the monopsonistic power of the purchasers and intermediaries" (Weisskoff 1992, 1537). Thus, despite the existence of agricultural cooperatives, the bulk of the benefits from agro-exports did not go to the hundreds of thousands of peasant families, but to intermediaries, multinational agribusiness, and the state. As perceived during the Stroessner period, collective organization among peasant farmers, other than in state-sponsored cooperatives, threatened the control big landowners had over communities and over the labor of smallholders (Lernoux 1982).

To counteract the drastic control measures implemented by the state and the growing level of inequality in rural Paraguay, the Catholic Church, in conjunction with students at the Catholic University of Asunción, promoted the establishment of cooperatives in the hinterland centered on grassroots communities known as Comunidades Eclesiásticas de Base (CEBs).* These were created in the poorest departments of eastern Paraguay and were the result of the experience of the Ligas Agrarias Cristianas (Christian Agrarian Leagues) initiated in the 1960s. By 1986 there were around 400 CEBs in the country with more than 15,000 members. Their main objectives were to achieve equitable land redistribution and gain access to credit and technical assistance for their members. In addition, they promoted community development such as the establishment of schools and health centers. These organizations, however, were perceived by the Stroessner regime as a "communist conspiracy" and were quickly destroyed

*CEBs, or church-based communities, emerged in Brazil and spread through other parts of Latin America in the 1960s, largely inspired by liberation theology with a focus on social justice and the Christian preferred option for the poor.

by the military through the murder of cooperative leaders (Lernoux 1982, 21). In fact, as Fals Borda aptly puts it in reference to the cooperative movement of the 1960s in other Latin American countries, "the very idea, if carried to its logical conclusions, could constitute a threat to a system that is based upon a completely different philosophy and way of life" (Fals Borda 1971, 15).

In sum, except for cases of large, technologically sophisticated agricultural cooperatives such as those of the Mennonites, sanctioned cooperatives in Paraguay did not make a real effort to instill the Rochdale principles of democratic cooperative membership, cooperative education, autonomy, and social justice. As Cohn et al. point out, "leadership and trust could not develop in that atmosphere. Meetings were considered conspiracies and a casual remark might lead to arrest" (2013, 97). Far from promoting cooperation, the legacy of the Stroessner regime within the smallholder agricultural sector was one of suspicion and distrust toward fellow farmers and those in power.

Throughout the Stroessner years, agricultural exports grew continuously and nearly tripled during the decade of the 1980s. By 1989, 96 percent of total exports consisted of livestock, agricultural, and forest products. The agricultural population increased by 134 percent, going from 835,000 in 1955 to nearly 2 million by 1988 (Carter et al. 1996). According to land tenure data based on the 1981 agricultural census, 36.2 percent of farmers had less than five hectares of land, while 80 percent of the farms controlled 12 percent of the land, and the largest 0.6 percent of the farms occupied 61 percent of the land (Weisskoff 1992).

THE TRANSITION TO DEMOCRACY AND NEW EXPECTATIONS

After the Stroessner regime, the country went through a period of political instability and recession exacerbated by Argentina's 2001 economic crisis, which impacted Paraguay's financial sector and regional commerce (Fernández Valdovino and Monge Naranjo 2004). According to Rodríguez Alcala and Otter (2008, 147), poverty rates increased from 32 percent in 1997 to 41 percent in 2003. The end of the Stroessner regime also initiated a period of transitional democracy in which elements of old dictatorship mixed with a new Paraguay that was trying to solidify democratic processes and strengthen its geopolitical position vis-à-vis its much stronger neighbors, Brazil and Argentina.

For small producers, this period of transitional democracy generated tremendous expectations, particularly after the election in 2008 of the former Catholic bishop Fernando Lugo, also known as the "bishop of the poor," ended 62 years of conservative rule. Small rural producers in Paraguay felt that, for the first time, they had an opportunity to participate in the definition of the new Paraguayan democracy and its integration into the global economy. New expectations, however, continue to clash with a rural reality that lacks basic infrastructure and retains elements of the old Paraguayan social structure, in which 48 percent of the total cultivated area belongs to large soybean farmers and private corporations. Moreover, the hope embodied in Lugo's presidency was shattered four years later when the Paraguayan conservative elite removed him from office through an impeachment process that most Latin American countries perceived as a coup d'état.

Thus, the collective struggle in Paraguay presents a number of contradictions. On the one hand, smallholder Paraguayan cooperatives share a fairly entrenched and indisputable view that someone else—international markets, the state, the industrial sector, or large wealthy agricultural producers—define the expectations and livelihood strategies of small farmers. On the other hand, by modifying those same expectations, cooperatives and their members are beginning to participate in the making of their own destiny. These contradictions are shown in the cooperatives presented in this section, as their leaders and members attempt in different ways to facilitate smallholder participation in larger movements that seek structural change by publicly challenging conditions that threaten members' subsistence.

GRASSROOTS ORGANIZING AS A KEY TO COOPERATIVE-LED CHANGE

Lessons from Rural Paraguay

BRIAN J. BURKE

LATIN AMERICAN PROGRESSIVES have often hoped that cooperatives might unite peasants in the pursuit of social, economic, and political transformation. Allied with land reform, collectivism, liberation theology, and leftist politics, cooperatives were supposed to provide an organizational form to support campesinos' political dynamism and the more general transition from semifeudal conditions to economic democracy. History has rarely borne out these hopes. As early as 1971, Orlando Fals Borda reported on the many roadblocks that turned cooperatives into mere palliatives that helped campesinos adjust to new forms of socioeconomic marginalization but not build more just societies (Fals Borda 1971; Fals Borda et al. 1976). In country after country, scholars have found similar stories.

In this chapter, I examine the case of a cooperative that served for almost three decades as a dynamic social, economic, and political force in Paraguay, one of Latin America's most unequal countries. By organizing small farmers into rural community committees, the cooperative created cohesion among a rural base that was intentionally fractured during the Stroessner dictatorship. The committees allowed it to deliver technical assistance and marketing services that significantly improved the socioeconomic conditions of small farmers and challenged some of the structural disadvantages that reproduced their marginalization. Unfortunately, a series of crises and changes within the cooperative have led to the decline of rural organization, mobilization, and participation and the reversal of much of this early progress. Here, I discuss the lessons that can be learned from this decline and the cooperative's efforts to reinvigorate its rural activities.

FROM GOLDEN AGE TO RURAL NEGLECT

Founded in 1971 by a small group of teachers and farmers who overcame fears of organizing during the Stroessner dictatorship, the Coronel Oviedo Cooperative for Agro-Industrial Production and Services is one of the largest and most successful cooperatives in Paraguay. The cooperative quickly accumulated a majority urban membership and has provided urban and rural services through much of the department of Caaguazú. Successful savings and loans programs have made the cooperative a major player in the regional economy. Up to 2005, it held almost $2 million in personal savings accounts; provided millions of dollars in loans each year for production, commerce, and consumption; and had loaned more than $39 million over the course of its history.

These financial services and production and marketing support have been extremely important in Caaguazú, an "eminently campesino" department that is the country's second most important agricultural region (Campos 2003; Casa de Cultura n.d.; Última Hora/DGEEC 2001). The department's population grew rapidly in the 1950s and '60s with expansion of the agricultural frontier and the "March to the East"—a mass migration from the central region to the eastern border promoted during the Stroessner regime—and continued to outpace the national growth rate until the mid-1980s. Today, two-thirds of its population resides in rural areas and more than 60 percent is engaged in the primary sector (agriculture, ranching, hunting and fishing, mining, and forestry). In the last thirty years, sugarcane and soy production expanded rapidly to complement traditional cash crops like corn, cotton, and tobacco. Ranching has also grown significantly, in concert with increases in large land holdings, inequality, and rural wage labor. Caaguazú is disproportionately important to the national agricultural sector: though accounting for only 11 percent of cultivated land and 9 percent of the national population, the department employs 14.8 percent of the nation's agricultural workers and farmers and produces 12.3 percent of the total value of national primary production (Última Hora/DGEEC 2001; Casa de Cultura n.d.).

This high productivity is accompanied by high population densities relative to the rest of the country and problems of rural poverty, soil erosion, and deforestation. Many families lack the 10 hectares deemed necessary for household self-sufficiency. In 2001, 61 percent of Caaguazú households lived below the poverty line, and the average household income reached only 49 percent of the national average (Última Hora/DGEEC 2001). Poverty rates surrounding the city of Coronel Oviedo were slightly better, with a 46 percent overall poverty rate and an average income 50 percent higher than in the department as a whole. The department had an unemployment rate of 18.9 percent. Although 67 percent of the population had completed some primary education, 14 percent had no formal education and 12 percent of adults were illiterate. More than 75 percent of residents lacked access to running water and sewage, though most had electricity in their homes. In this context of small landholdings, low levels of capital, ecological deterioration, and relatively high population densities, one might expect cooperatives to play a leading development role.

In the 1980s and '90s, the Coronel Oviedo Cooperative began to expand its rural development programs. Working through 57 rural community committees, the cooperative provided technical assistance for production, crop quality improvement, sophisticated marketing, and the export of a variety of crops. It also promoted significant participation by rural members who were isolated by poor communication and transportation infrastructure (Moran and Villalba 1989). Rural members met weekly in the committees to discuss cooperativism, learn about and provide feedback on cooperative programs, and plan monthly meetings with cooperative administrators about local needs and issues. They received regular training and frequent technical assistance. Significantly, the committees united, organized, and mobilized small farmers who were regularly ignored or repressed during the Stroessner dictatorship. One of the cooperative's first major rural development programs, a partnership with the Japanese International Cooperation Agency to export high-quality specialty fruits and vegetables, provided an important new income source when cotton and other traditional cash crops were underperforming. By 1998, 30 percent of the rural members were selling fruits and vegetables to Argentina and Uruguay, and a new market in Chile was soon to open. According to cooperative agricultural extension agents and members, this "golden era" was made possible by committees that facilitated communication, participation, and active engagement with rural members.

At the height of its rural activities, the Coronel Oviedo cooperative suffered two unexpected crises that caused rural membership to decline sharply, led to the dissolution of many rural committees, and threatened the cooperative's development program. The first occurred in 1996, when an attempt to restructure cooperative management and elect new, more professionally trained leaders sparked internal conflict. In the end, the leadership remained the same, but several highly qualified young agricultural technicians who had supported the change and who were deeply committed to participatory rural development were fired or resigned. As a whole, the institution recovered from this crisis fairly well, but the rural sector suffered. The cooperative lost the momentum it had gained through several years of grassroots mobilization and export-led development, and its advisory board even discussed eliminating agricultural programs altogether to focus exclusively on more economically efficient urban programs and rural lending.

The second crisis, of much greater consequence, resulted from a 1999 law that forgave all public debt to the agricultural sector (for more on this, see Burke and Piekielek 2011). Rural members of the Coronel Oviedo cooperative expected substantial benefits from this law, but the fiscally conservative cooperative had already repaid virtually all of its public debt and was therefore not obliged to forgive members' debts. Rural members saw their neighbors' debts vanish and felt cheated by the cooperative. The cooperative offered to renegotiate the terms of loans but failed to resolve misunderstandings about why members did not qualify for debt forgiveness. In Caagauzú, there is a long history of campesino mobilization (Campos 1993), including organizing by the Federación Nacional de Campesinos (FNC), a campesino political organization that helped negotiate debt forgiveness. FNC representatives encouraged Coronel Oveido members to protest by refusing to repay their loans to the cooperative. Approximately 500 of the cooperative's 1,200 rural members made the difficult decision to leave the organization that had been their most consistent support over the years. Many others stopped repaying their loans and were expelled from the cooperative.

Although new rural members continue to join the cooperative, the impact of the debt forgiveness law cannot be overstated. The loss of confidence, communication networks, and rural members and community leaders effectively ended farmers' organization through rural committees, reversed most of the gains made through the vegetable and fruit marketing project, and damaged the cooperative's ability to provide effective and efficient services in rural areas. The cooperative director called the debt forgiveness conflict the greatest challenge the cooperative has faced: "That is precisely where we fell. . . . It was our fault, but the debt forgiveness caused everything to disintegrate." In our interviews, long-term cooperative members regularly recounted the extent of cooperative assistance prior to the crises of the 1990s, using it as a point of comparison for the low levels of organization and interaction today. Many farmers felt left behind by the cooperative and said this was only made worse by the cooperative's all-urban advisory board. Cooperative staff and the board members themselves recognized that the board was ill-equipped to make informed decisions about rural issues. Farnan also described farmers' sense of betrayal and outrage at the cooperative as a "sea change" from the excitement and optimism she witnessed from 1996 to 1998 (Farnan 2007, 49). Because the cooperative has seemed to abandon rural development activities in favor of lending, the next generation of farmers and virtually all of the new members we interviewed are learning to see the cooperative as a bank rather than a leader in social organization and rural development.

In addition to the two crises mentioned above, the cooperative's financial success and expansion have contributed to this sense of neglect. The cooperative has expanded its zone of action based on its financial capacity but without regard for its capacity to provide personalized and labor-intensive technical assistance. Lack of transportation and a poorly maintained road infrastructure aggravate the problem, especially during the rainy season, when many rural roads are impassable. Despite its significant resources, the Coronel Oviedo cooperative does not have enough agricultural extension agents to provide the quality and duration of assistance that the region's small and medium-sized producers need. As a result, the technical assistants focus their efforts on those producers who have credit with the cooperative, and especially on making sure that the credit is repaid. While this is certainly important, it is not adequate for reinvigorating rural development. This not only limits rural members' ability to access services, but also limits their participation in cooperative decision making, exacerbating the underrepresentation of the agricultural sector in cooperative management.

With that said, the Coronel Oviedo cooperative remains an important resource for small and midsize farmers. It continues to provide some technical assistance, low-cost agricultural inputs, and health services, and the cooperative's credit lines are its most important service. While rural members face difficulties accessing many of the cooperative's urban-based services, they typically receive a disproportionate amount of the cooperative's credit. As multiple cooperative staff people informed us, "the urban members practically subsidize the rural ones." Cooperative loans are also highly favorable: as low as 12 percent for agriculture, compared with rates of 20 to 50 percent at commercial banks. Providing credit at favorable rates can make substantial contributions to rural members' livelihoods and to the sustainability of the cooperative itself. Rural members use loans for on-farm investments (land, machinery, inputs, storage facilities) and off-farm activities (stores, taxi cars, commercial photography equipment) and as family supports in the form of salary advances and education loans. By providing millions of dollars in loans each year for production, commerce, and consumption, as well as personal savings accounts totaling almost $2 million, the cooperative is a critical economic force in the region.

The importance of credit for farming and nonfarming households has enabled the Coronel Oviedo cooperative to use credit to retain old members, recruit new members, and maintain organizational stability. In response to the global economic recession at the end of the 2000s, they have increased their maximum loans from 10 billion to 40 billion guaranís. This has caused the cooperative to expand dramatically from 10,656 members in 2004 to 22,000 in 2010. Of these, 2,000 to 3,000 are farmers with an average landholding of 20 hectares, and another 5,000 are "mixed members," who farm as part of their livelihoods. The cooperative now works in nine districts in Caaguazú and two in the neighboring department of San Pedro, where it has opened a satellite office. New cooperative leaders—including the first new director in nearly 35 years—have made significant changes to the administrative structure, have facilitated member participation and benefits, and are now seeking ISO 9001/08 certification.

REINVIGORATING THE RURAL

During our first season of field research in 2005, cooperative leaders were already aware of the decline in the rural sector and the need to bring rural members on to the advisory council, to reinstitute rural community committees, and to develop more rural leaders. They initiated several projects to reinvigorate development activities, each of which combined agricultural and marketing assistance with community organizing. These projects also fit within a broader strategy of combining agro-industrial specialization with smallholder diversification. Agro-industry is meant to move Paraguayan farmers up the value chain so they can find bigger markets, better prices, and avoid competition with large, mechanized Brazilian farms, but it entails large capital costs and can generate substantial risks through dependencies on single crops and markets. Industrializing and specializing as a cooperative shifts some of these costs off the individual farmer, but farmers remain heavily invested in their specialty crop. In contrast, smallholder diversification tends to favor subsistence production over income generation and local over distant markets. Diverse smallholders are less dependent upon a single crop or market, but also less likely to earn substantial incomes.

The Coronel Oviedo cooperative attempts to provide farmers with higher incomes through specialization, industrialization, and links with distant markets while also building resilience through small-scale diversified production for local markets. As the former director of production explained, "The sporadic income [provided by agro-industry] is for development. The continuous income is for survival." In the last six years, the cooperative has established several resilience-building projects while devoting the bulk of its resources to the production of ethanol and sugar for export via organic and fair trade markets.

The *granja escuela* (farm school) project began in 2000 with nine rural households to demonstrate techniques for efficient and sustainable production. Cooperative leaders hoped that these model farms would be "poles of attraction" for other rural members, informal community centers where the cooperative could share strategies for converting family farms into efficient agribusinesses. The cooperative works with each family to develop a farm production plan that seeks to maximize productivity and ecological sustainability while balancing diversification with industrialization or the creation of high-value products. Twice a year the cooperative also invites the wives/mothers to the cooperative offices for a special program designed to provide instruction about family health and household finances, as well as to solicit their feedback. The cooperative is careful not to give anything to these families free of charge and not to provide special credit lines for them. It insists that each family have daily income from products such as milk, cheese, and eggs, which can be sold locally. The goal is for each family to serve as a realistic and replicable model for other farmers in their district. The program has been a modest but important success. By 2005, six of the nine families had already bought additional land with their increased earnings, and by 2010 the program had expanded to include 15 farms, each of which served about 20 local community members. Cooperative technical assistants visit the farm schools every two weeks to meet with all of the members.

To capitalize on the heterogeneous urban-rural membership of the cooperative and resolve the marketing problems faced by rural members, they also initiated twice-weekly farmers' markets in Coronel Oviedo. In 2005, cooperative leaders were also considering the construction of a cooperative supermarket to provide a permanent local market for all member farmers and benefits to urban consumers, but this idea was not pursued further. By 2010, four farmers' market groups consisting of campesinos living near the city had assumed primary responsibility for running markets on Saturdays, Tuesdays, and Wednesdays. They decide who can sell at the market and ensure that prices are fair. The NGO Alter Vida helped initiate the farm schools and farmers' markets, and also provided organic certification for sugarcane.

In addition, the cooperative has supported aquaculture, beekeeping, the production of goat's milk and related products, and orchards. The small-scale aquaculture project has proceeded through a committee of 32 farmers with a total of 150,000 fish. The beekeeping project developed through a broad partnership with government and nongovernmental organizations to prepare farmers to export honey. The goat's milk project began with 12 farmers and 200 animals in 2005 and has expanded to 43 households that are raising more than 4,500 goats. Working with an NGO, the cooperative identified and bred a brucellosis-resistant stock of goats and provided favorable credit to farmers who wanted to purchase animals, fencing, and other supplies. It also built an education center and purchased computers for farmers' children. The fruit orchards are a relatively new project in response to a government initiative to make Caaguazú a major fruit-producing region. The cooperative built a plant nursery with over 30,000 plants that it provides to members at cost, and it plans to construct a fruit-processing factory once the ethanol factory (discussed below) is paid for, to enable local producers to earn more for their fruit. These projects have succeeded—despite the obstacles of funding and human resources experienced by most cooperatives—because they have capitalized on alliances with the government and nongovernmental organizations and because they have involved intensive focus on relatively easy products with tested markets. Products like honey are especially safe because they can be saved until prices are high.

The cooperative's largest rural initiative, however, is the organic sugar and ethanol production project. Ethanol production was meant to sustain farmers during the three- to four-year transition to certified organic sugar for export.

It also responded to the cooperative's need to invest a significant amount of money, and to a rapidly growing domestic biofuel market. Approved in an extraordinary assembly in 2005, by 2010 the ethanol project was a controversial issue. In the words of one cooperative director, the project has passed through a profound crisis at the level of the administration and the cooperative members, and the technical assistance department will suffer as well. The stakes are higher than ever: as one cooperative director said, "The factory has to be our priority. It has to work optimally because of the debt that the cooperative took on for its construction and the crisis that it has generated with members."

What sparked such controversy? When members first approved a $2 million ethanol plant that would begin production in just over a year (by January 2007), hundreds of farmers signed up and began to modify their farm plans to include organic sugarcane. The cooperative, in turn, was supposed to move quickly on the factory so that farmers could receive a relatively high price for their new crops, but when we returned to Coronel Oviedo five years later, the factory was still only 70 percent completed. The cooperative's decision to double the plant's capacity compounded financing problems and led to an initial delay that troubled farmers. In response, farmers demanded a review of the entire project that concluded favorably but postponed construction further. All told, these expansions and setbacks caused costs to balloon from between $1 and $2 million to between $10 and $12 million. The cooperative had organized farmers into 34 sugarcane production committees, expanded sugar production with a preferred variety, provided loans to enable farmers to change their production systems, and began the organic transition and certification process, but many farmers felt cheated. The cooperative did not keep up its end of the bargain, and the farmers suffered through multiple years of low profits; many farmed sugarcane at a net loss.

Today, the cooperative is working with 500 farmers responsible for 1,500 hectares of sugarcane, about half of the plant's total capacity. It plans for the project to benefit 1,000 small farmers, employ 136 people directly in the factory, and provide indirect benefits to 8,000 to 10,000 people in total. Cooperative employees said that the plant would be finished before the end of 2010, but as one grower noted, it is very hard to believe this now, after five years of waiting and losing money on sugar production. Moreover, the cooperative has recently learned that they need to balance their cane crop with at least 20 percent early-yielding varieties. After taking loans and spending large sums of money to plant sugar, members are resistant to another change. The cooperative is distributing early-variety seeds for free and has agreed to refinance all of the sugar production loans that it made.

To make matters worse, the cooperative's intensive dedication to the sugarcane project has hurt other initiatives. Some members feel that the farm schools have been shunted aside, and others agreed that for the last five years nearly everything else has been neglected. Only two extension agents have remained to work on the smaller livelihood diversification projects, and this staffing decline is not likely to improve. Sugar production and the ethanol plant will continue to absorb large amounts of the extension agents' time. The ethanol plant may pay off nicely, but it could easily become a third blow to rural members' trust in the cooperative.

CONCLUSIONS: GRASSROOTS ORGANIZING AND AGRO-INDUSTRIAL PRODUCTION

The recent history of the Coronel Oviedo cooperative has been marked by ups and downs. Many members' household economies continue to improve as a result of the help they receive from the cooperative. Support for production, marketing, and especially credit—which has been a key constraint on rural development throughout the cooperative's history—have enabled the majority of members to acquire more land and increase and diversify their production, rather than selling land, shifting to rural wage labor, and migrating to the cities as many other Paraguayan peasants have done. But, as one cooperative director noted, "The cooperative has still not recovered from the crisis of 1998 and has not managed to refocus." There is still much work to be done if the cooperative is to become an agent of change promoting a new, more just society. Rural mobilization will need to be part of that effort.

Much of the cooperative's early success was due to the organization of campesinos in active community committees. In alignment with core cooperative principles, these committees united small farmers in the pursuit of their own development goals and permitted economic dynamism that would have been impossible separately. Over the last six years, the cooperative has tried to reestablish these committees in tandem with their new development projects. They have created 34 sugarcane committees, 15 farm school committees, 4 farmers' market groups, and 3 committees for producers of goat's milk, honey, and fish. One of the committees consists exclusively of women, after they decided to remove the men from their group for being disruptive. Several of these committees demonstrate the importance of organization: the sugarcane committees set their own rules and have held the cooperative accountable to its rural base; the farm school committees permit the flow of information across communities; and several others have initiated their own projects, independent of the cooperative, in partnership with government agencies and NGOs. The farmers' market organizers are also nearly autonomous. By putting these dynamics in place, the cooperative is creating the foundation for a self-organizing rural society that is no longer dependent on cooperatives, banks, and the government but can now work to hold those entities accountable. By contrast, many of the problems with the sugarcane project are due to the cooperative's effort to force a highly technical and capital-intensive project onto producers with little faith in their capabilities.

Strong rural social organization can also benefit the cooperative and rural members by permitting effective and efficient outreach where geography and infrastructure hamper regular contact with cooperative extension agents. Perhaps most important, community committees can facilitate communication between the cooperative and different segments of Paraguay's increasingly heterogeneous rural society. As our survey results revealed, not all rural dwellers and not even all small farmers are the same. A farmer with 10 children who borrows 5 hectares of land from his in-laws has very different needs than a farmer with 2 university-educated children who works 150 hectares with machines. A family of two rural teachers with a small garden parcel will have a third set of needs and interests. In order to effectively serve such disparate groups, cooperatives need to maintain strong participation by and communication with their members. They also need to continually investigate the self-expressed needs of their members. This approach might even enable Coronel Oviedo to work with the poorest farmers in the region, something that cooperatives often fail to do.

In the process of reestablishing these committees, however, cooperative directors and rural members have learned that it is far easier to undo social organization and active participation than to restore them. Several small farmers and cooperative employees noted that many of the committees are weak and confidence in the cooperative is wearing thin, especially due to the difficulties with the ethanol plant. Participation in cooperative decision making continues to be a challenge due to the cost and difficulty of travel to the city, the timing of cooperative meetings, and in some cases an attitude of justified indifference on behalf of campesinos. As one dairy farmer told us, "I am very ill informed about the cooperative because the cooperatives leaders don't understand what's happening with its rural members." Five years after the cooperative leaders said that they understood their own deficiencies and wanted more rural representation, they still faced many of the same problems.

If the Coronel Oviedo Cooperative wants to be more than a bank, if they want to return to being a dynamic cooperative with broad impacts on the social development of rural Paraguay, they will need to return to emphasizing the *social* aspects of cooperativism—including solidarity, participation, mutual support, and self-organization—alongside the *productive* and *marketing* aspects of rural development. This is an important lesson for other large rural cooperatives as well. Agro-industrial production may help promote trust, solidarity, and participation if it can generate economic gains efficiently and without jeopardizing farmers' already vulnerable household economies. However, when the path to agro-industry requires major changes to farm systems and when it meanders through lofty hopes and unfulfilled promises, the damage is likely to be both economic and social.

COLLECTIVE CHALLENGES AND THE FAIR TRADE ALTERNATIVE

The Manduvirá Cooperative

MARCELA VÁSQUEZ-LEÓN AND
ANDRÉS GONZÁLEZ AGUILERA

AS YOU ENTER THE TOWN OF ARROYOS Y ESTEROS, in Paraguay's department of Cordillera, you can't miss a sign on the main highway that greets visitors: "Welcome to the organic valley of Arroyos y Esteros. The world's organic sugar cradle." The sign is appropriately jubilant, but the region's path to organic success has been complicated. Much of their success has depended on the Cooperativa de Producción Agroindustrial Manduvirá Ltda. and its ability to channel individual initiative into collective action.*

In a period of fewer than 15 years, the Manduvirá cooperative transformed itself from a bankrupt agricultural cooperative to one of the largest global producers and exporters of organic sugar for international fair trade markets. By following an alternative development model based on principles of environmental sustainability and social equity, the Manduvirá cooperative has secured a space for its growing number of producers in the global economy. It has done so by building alliances with international buyers and nongovernmental organizations while managing to maintain independence from state bureaucracies and Paraguay's large industrial private sector.

Cooperative leaders, with strong support from members, have been able to use organic certification and fair trade to enter global markets, break the

*This chapter is an updated and modified version of a previously published journal article (Vásquez-León 2010).

abusive monopoly of the local sugar mill, and achieve greater control over the value chain. In less than a decade the cooperative went from dependence on the abusive practices of the sugar mill, called OTISA (Oficina Técnica Industrial, S.A.), which paid below-market prices for the cooperative's sugarcane, to renting the use of Censi & Pirota, a sugar mill 90 kilometers south of town, and creating their own brand of sugar. Today all rural cooperative members are certified organic producers and their sugarcane is processed by the cooperative's recently inaugurated sugar mill. The cooperative exports and markets its sugar directly to buyers all over the world. At the same time, it returns earnings to cooperative members in the form of direct cash and through social services for members and the community. This chapter examines the strategies of the Manduvirá cooperative to empower marginalized producers by asserting independence vis-à-vis the rural elite and global competitors while achieving autonomy, self-governance, and long-term stability. Examining the strategies of its leaders and members reveals that participation in global markets requires active involvement in debates on globalization and democracy, the redefinition of production processes to satisfy external markets, and recognition of the needs of small producers to make a reasonable living and define their own development path.

THE COOPERATIVE: BACKGROUND AND SETTING

Manduvirá is a mixed-service cooperative that serves 950 small farmers, who typically own 3 to 5 hectares of land, as well as 750 urban residents of the town of Arroyos y Esteros (DGCE 2002). Arroyos y Esteros is essentially agrarian, with the town's population of 4,300 people dwarfed by the rural population of 22,000. However, soils are fragile and fertility is low as a result of historically high population pressures and poor soil management. The region has a long tradition of sugarcane production for household consumption and local markets in the form of sugarcane syrup and alcohol. In the past, the region also depended on the production of pineapple and bananas, crops that became unfeasible due to soil degradation and unstable markets. Lack of soil productivity was one of the main factors that contributed to the out-migration of entire communities during the 1970s, when Paraguay's military dictator Alfredo Stroessner undertook a massive internal colonization project known as the Settlement of the Northern Axis. Although this out-migration reduced pressure on the overpopulated and overused lands in the central departments, including Cordillera, it also left the rural district of Arroyos y Esteros essentially abandoned by the state.

Like other rural districts where small farmers predominate, and despite its location only 54 kilometers from Asunción, Arroyos y Esteros remained isolated through much of its history. Communication and transportation difficulties limited farmers' access to modern agricultural technologies and inputs, including chemical fertilizers and pesticides. Thus, productivity remained low and farming was, and still is, mostly organic. The region's isolation, however, ended as the twenty-first century began. A new highway connecting Arroyos y Esteros to Asunción, which also traverses eastern Paraguay, has made the town a high-transit zone for cargo trucks and passenger vehicles. Roadside businesses have multiplied, and faster communication with Asunción has opened the possibility of accessing the largest urban market for agricultural produce. Whereas in the past, organic production signified poverty and marginalization, today it facilitates a new social identity of environmental stewardship, social justice, and global interconnection. Since 2000, sugarcane has became the most important income-generating crop in Arroyos y Esteros. In addition, farmers also cultivate fruits and vegetables for household consumption and local markets.

COOPERATIVE HISTORY: FROM FAILURE TO UNPRECEDENTED SUCCESS

Manduvirá was established during the military dictatorship in 1975, when 39 schoolteachers and farmers decided to participate in a national plan to create savings and credit cooperatives under close scrutiny by the state. During its first years, the cooperative became the only source of affordable credit and membership grew quickly, reaching over 400 members by the beginning of the 1980s. To expand credit services, the cooperative borrowed from the National Federation of Savings and Credit Cooperatives, CREDICOOP, created by the state in 1979.

By the mid-1980s, however, the cooperative was facing financial crisis. Easy access to CREDICOOP loans led to the issuing of credit without regard for members' repayment capabilities. As one of the original members explains, "We lived through a long period of crisis from 1983 to 1989. Loans were authorized based on political criteria and debts accumulated. Cooperative officials were irresponsible and illegal transactions, as well as the embezzlement of funds, became the normal state of affairs." By 1985, with an accumulated debt-to-asset ratio of 105 percent, the cooperative went bankrupt. Membership dropped, and a new manager was assigned by the state to recover and liquidate cooperative assets. With the end of the military dictatorship in 1989, expectations of new economic and political opportunities encouraged the cooperative's board of directors to try to salvage the organization. After successfully renegotiating its debt, the cooperative began to market sugarcane syrup from members, who processed it in the 150 home-based artisanal sugar mills located throughout the district. Manduvirá transformed from just a savings and credit cooperative to one that would include an agro-industrial component, and sugarcane syrup became its most important source of earnings until the mid-1990s, when prices began to decline.

In 1992, the Fairtrade Labeling Organization (FLO),[†] an international NGO that sets fair trade and organic certification standards, visited Arroyos y Esteros. FLO's marketing pitch persuaded one of Paraguay's largest private-sector industrialists, who was also the president of the Paraguayan Industry Confederation, to purchase OTISA, a local sugar mill and distillery that had been built in the 1950s. The new owner retrofitted the mill to process organic sugar, and the mill was certified by FLO in 1997.[‡] As sugar syrup prices plummeted, artisanal sugar mills went into disuse, and the cooperative was forced to discontinue sales of sugar syrup. Cooperative members turned to OTISA to market their products, and the cooperative reverted to a savings and credit organization, continuing to offer credit for agricultural purposes and to rent machinery for land clearing, soil preparation, and sugarcane.

The relationship between producers and OTISA was filled with tension from the beginning, and, because farmers entered into direct contracts with OTISA, the cooperative had no power to negotiate on behalf of its members.

[†]FLO mediates the fair trade market. According to its website, http://www.fairtrade.net, the Fair Trade label guarantees that products conform to fair trade socioeconomic and environmental standards.

[‡]Certification requires third-party verification. It consists of an initial inspection of facilities and the production process, a review of the industry's regulations and production standards, and inspection of all certified producers within the cooperative. If certification is granted, an organic label is awarded. The certification process may take between two and three years as producers shift from conventional to organic methods.

As one of the founding members said, "At that time the cooperative didn't intervene, didn't have an opinion, farmers were really on their own." Despite FLO's guarantee that OTISA's sugar conformed to fair trade standards and despite OTISA's claims of alleviating poverty in this "very isolated region . . . with no communication with the rest of the world" (OTISA 2005), prices paid by OTISA to producers were at least 25 percent lower than those paid by other refineries ($12 per ton as opposed to $16 at other mills). The only immediate benefit for the producer was an assured buyer that was conveniently located. In the longer term, producers could look forward to the benefit of the fair trade premium, a bonus paid on top of an agreed minimum price, which amounts to $60 per ton for conventional sugar and an additional $20 per ton for certified organic sugar. OTISA, however, often kept most of the premium.

This created a dilemma for FLO. Because their partner showed little concern for the rights of local producers, "fair" trade was re-creating old hierarchical relations of power instead of fostering social justice. However, sugarcane required processing facilities that no small Paraguayan smallholders could afford.

UNDERSTANDING THE COOPERATIVE ADVANTAGE AND THE NEED FOR AUTONOMY

Despite their rough entry into fair trade, the small producers of the Manduvirá cooperative have achieved substantial gains. One of the greatest trials has been changing the balance of power in relation to OTISA. In the late 1990s, producers began to meet in local agricultural committees, which were organized by family groups from different *compañias* (smaller rural units that form a district). Meetings dealt with social, political, and economic issues relevant to particular localities, including farmers' dissatisfaction with OTISA. Agricultural committees, however, were isolated from one another and, as one member put it, "We were complaining to each other all the time and getting our anger out on our families, feeling that we were not getting anywhere." Committee leaders began to identify the cooperative as a possible unifying institution that could make their demands heard and, ultimately, break OTISA's monopsony control.

Representatives from various local committees approached the cooperative's manager. As one committee leader explained,

> We went to his house and invited him [to represent us], and we told him about our idea. And he said yes, but that we had to help him because he didn't know all

> the producers. And we said that we needed the cooperative to lead our struggle and we would invite every producer to participate. So we worked for five or six months, inviting producers, talking to them, knocking on people's doors, one by one, convincing everyone that if we acted together we had a better chance. People were afraid of OTISA, they were afraid of everything, of the government; people had been terrorized for so many years. But all the time, the manager of the cooperative and the board of directors were at our side.

After several months of community organizing, a meeting was set to negotiate with OTISA's owner at the cooperative. As the manager remembered, "There were more than five hundred producers at the meeting; it was the first time in history. We started negotiating but [OTISA's owner] didn't take us seriously." The meeting ended without a resolution, but the cooperative members continued to mobilize. The cooperative manager continued, "When the sugar mill gave producers the order to start cutting the cane, no one cut. . . . Despite fears, we decided to wait for several days until the mill agreed to raise prices by twelve thousand guaranís.[§] That was very significant; people realized that in an organized way we could accomplish things."

This modest gain made members aware of the strategic importance of collective action. Producers began selling their sugarcane to OTISA via the cooperative, and in 1999 the cooperative was able to gain control over the distribution of the fair trade premium.

The second step towards breaking OTISA's monopsony was through the creation of viable competition. In 2002, cooperative officials successfully negotiated the reopening of Censi & Pirota, a small sugar mill located 70 kilometers south of Arroyos y Esteros, by guaranteeing the delivery of a specified volume of sugarcane. Despite the greater distance and lack of certification, the new mill offered better prices and was willing to absorb transportation costs. Even though OTISA raised prices once again, most producers shifted to the new mill. In the words of a woman producer, "We knew that OTISA exploited us and that with the cooperative we were going places. In the cooperative we could talk and argue and say what we thought and we were not afraid anymore. That was worth the risk of going against the powerful." This shift was further encouraged when Censi & Pirota obtained FLO certification, allowing it to pay better prices for certified sugarcane. Without producers, OTISA has had to buy land and produce its own sugarcane.

[§]One U.S. dollar equals approximately 5,000 guaranís.

For the 10 years following this battle, cooperative members processed their sugar in the Censi & Pirota mill, increasing production of their own brand, Azucar Manduvirá, and selling directly to international buyers. Through these years, the cooperative's vision of autonomy and empowerment of its producers continued to grow, and in 2011 the cooperative embarked on what a few years earlier had seemed like a faraway dream: the construction of its own sugar mill. The cooperative decided to build its own mill partly to gain independence and lower production costs, but also because Cenci & Pirota were unable to process the high amounts of sugar provided by members. This vision of autonomy and empowerment started to form after the cooperative won its first battle with OTISA, but it became stronger and better defined with each successful step at controlling the value chain.

In the spring of 2014, with the financial backing of long-term international buyers, credit from the Inter-American Development Bank, and the technical support of long-term volunteer consultants, the cooperative inaugurated its own sugar mill in the outskirts of Arroyos y Esteros. This $15 million project is "the world's first producer-owned Fairtrade organic sugar mill" (Fairtrade International 2014). The new mill, which started production in 2014, has the capacity to process 200,000 metric tons of sugarcane and produce 20,000 metric tons of organic sugar per year.

GROWTH AND THE STRENGTHENING OF COOPERATIVE VALUES

The new sugar mill required the cooperative to hire and train 170 new employees. To increase production in a sustainable way, the cooperative also hired 50 employees to work with producer-members in different capacities. Most of those hired are sons and daughters of cooperative members, and many of those are returning migrants who had left the area in search of work opportunities in Asunción or Argentina. The hiring and training of young people from the community reflects a conscious commitment made by the cooperative since it hired its first agricultural extension agent in the mid-2000s. This extension agent is a producer's daughter who received funding from the cooperative to get a bachelor's degree in agricultural sciences at a Costa Rican university and to specialize in organic farming. Today she heads the cooperative's agricultural department, overseeing organic certification programs that also include vegetable production for household consumption and sale in regional markets.

An emphasis on capacity building has been at the core of the cooperative's strategy to grow without losing sight of cooperative principles. To prepare for the opening of the sugar mill, the cooperative funded several training programs for new employees and new member-producers. With a grant from the Inter-American Development Bank, the cooperative hired technical experts from Cuba, Costa Rica, the Czech Republic, and Brazil to train new employees. Trainings, however, were not limited to learning about sugar processing and working and managing the mill. In addition, the cooperative provided workshops on cooperativism, fair trade, organic certification, food security, quality control, and financial management. The cooperative invested $100,000 to train its new workforce, paying minimum wage throughout the four-month training period.

INTERNATIONAL SUGAR MARKETS: FAIR TRADE VERSUS FREE TRADE

Despite the challenges related to organic certification, choosing a fair trade strategy has been fundamental to the cooperative's success, as small producers have no realistic possibility of competing in the highly competitive international free market for conventional sugar. Paraguay's sugar industry, which is constituted by the country's largest producers, provides only about 1 percent of the sugar that enters MERCOSUR and is in direct competition with Brazil, the largest exporter of sugar in the world. World sugar prices are also highly volatile and have been on a downward trend since the 1980s (Illovo Sugar 2007).

In contrast, Paraguay has become an important producer of organic sugar for fair trade markets. Sugar, however, is only a sideline commodity, with an approximate total production of 190,000 tons as predicted by the International Sugar Organization for 2005 (Gudoshnikov 2001). The share of organic sugar for fair trade markets is even smaller, though it has shown significant growth according to several measures: European demand expanded by 120 percent in 2003, from 650 tons to 1,430 tons (Transfair USA 2009); Fairtrade International's sugar sales increased from 16,523 tons in 2007 to 211,600 tons in 2013; and FLO has significantly increased the number of producer associations it works with, from 16 associations in 8 countries in 2007 to 100 associations in 19 countries only 6 years later. The Manduvirá cooperative is one of the lead producers of organic certified sugar, with an average production of 7,500 MT. The cooperative, however, is now expecting a production of 12,000 MT for 2015 to meet increasing demand, and it expects to reach the sugar mill's total capacity of 20,000 MT. The cooperative currently exports sugar to 24 countries, including most European countries plus Brazil, Peru, Chile, Ecuador, Uruguay, Mexico, Canada, the U.S., Australia, New Zealand, Hong Kong, and South Korea. It is now also working through other certifying agencies like Trans Fair and Trans Aid. Demand comes not only from anonymous consumers but also from food producers, supermarkets, and cacao cooperatives that directly contract the cooperative to supply them with fair trade sugar for use in their organic chocolates, ice cream, jams, and baked goods.

Price differentials also make organic sugar more attractive to producers. According to the International Sugar Organization, organic sugar commands about one and a half to two and a half times the factory price of conventional sugar (Gudoshnikov 2001). The president of Paraguay's largest sugar mill estimates that, whereas conventionally produced sugar fetches about $260 per ton, organic sugar sells for about $330 per ton. In his opinion, the organic sugar industry "is a way to compete on the international markets against Brazil's cheap sugar" (Sciscioli 2005).

For cooperative members, production for fair trade has increased and stabilized annual household incomes. According to the Fairtrade Foundation, the fair trade minimum price guarantees "sustainable production" costs and acts as a safety net for farmers when world market prices fall below a sustainable level. When the market price is higher than the fair trade minimum, buyers must pay the market price. Also, the cooperative can request partial prepayment of the contract, and it is in fact exempt from prefinance requirements (FLO 2009). This assures the cash flow to pay farmers when they deliver their crop.

In addition, the premium has steadily increased as more farmers become certified. For the 2007–2008 season, for example, when 60 percent of the producers had received fair trade and organic certification, the cooperative received a total premium of $600,000 for the production of 7,500 tons of sugar. Half of this amount goes directly to the producer, yielding an average of $360 per producer during the 2007–2008 season. According to producers, adding the premium to higher prices per ton yielded a 30 to 40 percent price increase compared with when OTISA was the only buyer. The other half of the premium, following fair trade requirements, is invested in community projects, allowing the cooperative to extend solidarity networks beyond the cooperative. In 2004, for example, money from the premium was used to construct a community health care center.

MANDUVIRÁ TODAY AND FUTURE CHALLENGES

The rapid success of Manduvirá has, in a little over a decade, turned the cooperative from a futile organization to a highly dynamic one that is constantly changing as it seeks to expand its economic activities, provide benefits to a rapidly growing membership and the community in which it resides, and obtain full control over the production and processing of its members' sugarcane. With the new sugar mill, the cooperative is enjoying higher prices and paying lower transport costs. Also, cooperative farmers are 100 percent fair trade and organic certified. And by creating new employment opportunities for community members, the cooperative is ensuring that benefits remain in the community. Farmers also expect that the quality of the sugar will increase, since the new mill allows freshly cut cane to be processed immediately.

Success, however, has brought new challenges as well as new strategies to deal with those challenges. Despite growing demand, fair trade organic producers around the world continue to increase, and the end result in terms of markets is unknown. This problem magnifies when one considers the fact that overreliance on sugar led to an initial decline in agricultural diversification. The cooperative has tried to address the problem by looking for new markets, such as sesame seed exports to Japan, but, so far, market entry has been difficult. It has been more successful in consolidating strategic alliances with external organizations interested in sustainable agriculture. Alter Vida, a Paraguayan nongovernmental organization whose primary objective is to "promote agroecology as part of the Paraguayan culture," provided technical assistance to certified producers for the establishment of organic gardens for household consumption and sale in an organic farmers' market. The amount of organic produce has increased substantially; today every household in the cooperative is producing for consumption, and regional produce markets have become a regular event. With these initiatives the cooperative has already placed food security as a priority.

As a long-term strategy, the cooperative continues to build capacity by investing in education. Since the mid-2000s, the cooperative provides school supplies, computer classes, and English and music lessons, and has an internship program for high school students to learn about business management, accounting, marketing, and agronomy. The cooperative's agricultural department has expanded, offering a variety of opportunities for farmers. As one member explained, "We get workshops, we learn new ways of doing things and we mix them with the old ways, we feel more secure that we can take a step forward." The cooperative is continuously training young men and women from the community to learn new skills and become part of the larger project. One 18-year-old high school graduate explained, "The truth, I feel proud, the cooperative is a big family. . . . We are not just producers, we are also manufacturers of sugar, marketing agents, agronomists. Before, the only thing was to be a sugarcane farmer to be exploited by the rich. Were it not for the cooperative I would be looking for work in Argentina." Even though the cooperative has not had women in top leadership positions, the cooperative's extension agent today is a young woman whose parents are cooperative members. Today her leadership is fundamental to the inclusion of women farmers in cooperative programs, including support for subsistence farming as well as for farmers' markets, where women tend to be the ones who sell.

Up until 2013, the cooperative had an open door policy, but with all members acquiring organic certification, new members entering the cooperative, and greater benefits accruing, the cooperative went into what the manager calls a "growth crisis." To deal with the crisis, cooperative managers have decided to halt the entrance of new members until the cooperative stabilizes. However, the cooperative already has a good foundation for continuous growth, which underscores the need for governance and transparency. Besides holding regular meetings where producers are able to voice their opinions, the cooperative has created a complex decentralized structure in which local producers in the different compañias take turns coordinating the harvest and transportation from local collection sites to the sugar mill. Managing buyers has also become an overwhelming task: e-mails arrive from all over the world in different languages, challenging cooperative leaders to learn English and become computer literate.

A barrier to growth that seemed significant was lack of support from Paraguayan state and industrial sectors. In Paraguay, Manduvirá has generated distrust because it has used the cooperative model to break a cultural and political economic paradigm. Without the permission, investment capital, and marketing assistance of the state-industrial elite, the cooperative has successfully linked small-scale, resource-poor producers with international consumers, providing the means by which these producers, rather than the traditional elites, fully benefit from their labor. They have also succeeded in taking control of the value-chain without destroying the character of small-scale agriculture. Instead they have strengthened solidarity networks. In the words of one small producer: "We have embarked on a great adventure, and our most important objective is to keep our institution on the right track, to find new markets, achieve greater benefits for our families, and we are getting there; health, education, autonomy, and all the things that are important to us."

This is one of the most important lessons that Maduvirá, its leaders, and its members have to offer. This cooperative is successfully assembling and representing the voices of marginal people in discussions of globalization. By defying the status quo and stubbornly maintaining a path of self-definition and empowerment, this cooperative illustrates how grassroots development can engender a culture of solidarity and a sense of shared purpose that goes beyond the local. The cooperative is having a global impact on conversations regarding alternative markets and the participation of small producers in international markets. Manduvirá, as a formally recognized collective organization, has become an avenue through which marginal producers find new ways to construct an identity while making a reasonable living and exercising a certain level of control over the conditions in which they work.

Despite difficulties, the future looks promising. The various gains that the cooperative has made in such a short time have led to a widespread feeling of solidarity and shared purpose, as cooperative leaders and producers solidify a common strategy that is compatible with the cooperative's mission of social justice. Even though overreliance on sugarcane is a risk that Manduvirá producers have taken and that they know they need to address, the key for continued success lies in the cooperative's ability to figure out how to address its new challenges with the same sensitivity and attention to the grassroots level with which it addressed the sugar monopsony.

IS MONOCULTURE A VIABLE STRATEGY?

The Case of Guayaibi Unido

RODRIGO RENTERÍA-VALENCIA

"Bananas are a type of fruit that shows everything in their skin; one can take a look at a banana and determine in which step of the production process it was damaged."

—MAURO SILVA, TROPICAL FRUITS QUALITY STANDARDS SUPERVISOR, BUENOS AIRES CENTRAL MARKET

A CRITICAL ASPECT OF THE ANALYSIS OF small-scale grassroots cooperatives consists of documenting and explaining the strategies that producers pursue to create economic niches within transnational landscapes that are otherwise dominated by larger organizations with stronger political, economic, and technological capital. A common strategy emphasizes the resilience that results from diversification (see Piekielek and Finan, and Vásquez-León in this volume). Yet many smallholder cooperatives do not diversify their production and opt instead for a monoculture approach. This chapter sheds light on the social processes responsible for this persistent bet on monoculture (rather than diversification).

A central argument behind this analysis proposes that a number of social considerations (such as social prestige) may play a fundamental role in determining the internal drivers of a cooperative. This chapter thus aims to incorporate a larger anthropological consideration of social domains that may prove crucial in determining the goals and strategies followed by small agricultural cooperatives. To illustrate this, I document the achievements and perils of the Paraguayan cooperative Guayaibi Unido in its effort to export bananas to Argentina. Paradoxically, while the bananas produced in the region of Guayaibi are the object of pride among their producers, in Argentina bananas are considered a second-quality product—often consumed by the lower socioeconomic classes. Yet the very act of exporting these bananas to Argentina provokes in many producers a tangible sense of prestige that reinforces the will to continue exporting their product despite attendant adversities.

GUAYAIBI UNIDO: THE SETTING

The cooperative Guayaibi Unido is located in the region of Guayaibi, 250 kilometers to the north of Asunción, in the department of San Pedro, Paraguay's most contentious political state. San Pedro has a highly skewed land tenure system consisting of soy farmers and ranchers with between 5,000 and 20,000 hectares of land, a smallholder sector (about 40 percent of producers) with 5 hectares of land or less, and a considerable number of landless peasants who, although not accounted for in census data, are highly visible through conflicts with large landowners (Vásquez-León 2010). Up to the mid-1980s, the Guayaibi region remained isolated. Despite its 14,000 residents, the road connecting Guayaibi to the closest urban center, the city of Coronel Oviedo 120 kilometers away, was only built in the second half of the 1980s. Basic services such as water and electricity began to be provided in the 1990s, but residents of the communities around the cooperative only gained access to running water after 2000 through the aid of a nongovernmental organization. San Pedro is the main banana-producing region in Paraguay, where small farmers dedicate about 2,000 hectares of land to banana production. The cooperative's 80 farmer-members have slightly more than 200 hectares in banana production. On average these producers cultivate 10 hectares of diverse crops, including banana and pineapple, for market sale, and cassava, beans, and corn for household consumption.

Shortly after its founding in 1999, Guayaibi Unido managed to enter the Argentine market through a strategic alliance with the Central Paraguaya de Cooperativas (CEPACOOP), the only central cooperative in Paraguay that groups smallholder cooperatives, including six other small fruit and vegetable cooperatives (see Echagüe, this volume, for more details on the institutional context for Paraguayan cooperativism). Through the intermediation of CEPACOOP, these small cooperatives were able to expand their markets beyond the Central Market of Asunción and gain access to MERCOSUR countries. Selling bananas in the international market, however, requires that the cooperative follow strict quality standards, since it faces harsh competition from larger and more technologically sophisticated banana producers who also serve MERCOSUR consumers. This puts enormous pressure on cooperative members to adopt new production and quality-control measures.

Although Guayaibi Unido membership represents two-thirds of local banana producers, the cooperative is responsible for only 20 percent of banana exports from the area. Many producers are not able to place their product in the international market, because, as Vásquez-León points out, "even though everyone in Guayaibi appears to want to export, few have an understanding of who determines quality standards, why, and what must be done to meet those standards" (2010, 55). Despite the difficulties, the cooperative continues to focus on exports to Argentina. What explains this persistence, and what are the costs?

This question is important because it places Guayaibi Unido in a bind where expanding exports becomes a source of exclusion within the membership and thus affects the essential solidarity and reciprocity that characterize a successful cooperative. To understand this dilemma and its consequences, we must consider larger social considerations around prestige.

HISTORICAL BACKGROUND

The first wave of farmers to settle in the region of Guayaibi arrived in the early 1970s as part of the rural colonization project implemented during the military dictatorship (1954–1989) of Alfredo Stroessner. Despite government promises, few social or structural supports were granted to these early peasant migrants. There were no production loans, technical training, or infrastructural improvements such as paved roads or electricity. These meager conditions forced farmers to focus on subsistence agriculture. It was not until marketing agents arrived in the zone in 1977 that farmers were able to start selling products like cotton, peanuts, beans, soybeans, and, later on, pineapple and bananas. The lack of public support created a strong dependency on market intermediaries, who soon became the only source of financing for local producers.

Seeking alternatives to achieve some level of independence, in the early 1980s a group of producers in Guayaibi organized Comités Agrários, or agrarian committees. These committees, lacking any type of legal recognition, where nonetheless encouraged by the Ministry of Agriculture and Ranching (MAG) as a way to facilitate access to credit and enable producers to market their products without an undue presence of market intermediaries. With the end of Stroessner's regime in 1989, the MAG attempted to restructure the agricultural sector by consolidating already existing agrarian committees into larger units. In this context, in 1990, the more than 190 committees in Guayaibi fused into a single entity, forming the Asociación de Productores de Banana y Piña del Norte (Association of Banana and Pineapple Producers of the North). As an association, these producers were able to obtain a sales stand in Asunción's Central Market. At the same time, much needed improvements in infrastructure were finally introduced in the region—over the following decade access to water, electricity, and a paved road was consolidated. Today, the highway from San Pedro to Asunción connects the region to important domestic and international markets.

It soon became evident to cooperative members that legal recognition was imperative to access production loans. Thus, in 1997, the cooperative Guayaibi Poty registered as an official cooperative sanctioned by the Instituto Nacional de Cooperativismo (National Institute of Cooperativism), or INCOOP. However, the almost immediate emergence of internal conflicts provoked the expulsion of a section of its members, who readily formed the cooperative Guayaibi Unido. After obtaining legal recognition in 1999, Guayaibi Unido received a loan from the government's Fondo de Desarrollo Campesino (Peasant Development Fund), or FDC, which allowed it to expand banana and pineapple production and hire an agronomist and an accountant. The FDC's Technical Assistance Unit, funded by the International Fund for Agriculture, provided marketing assistance and leadership training from 2000 to 2005. The loan and training convinced members that the Guayaibi Unido leaders were capable of networking at different levels in order to develop a more viable commercial strategy. The technical assistance program was designed to develop a marketing strategy implemented under the umbrella of CEPACOOP. Midway through the project, however, a group of discontented cooperative members decided to separate and create the San Gabriel cooperative. They argued that Guayaibi Unido was favoring only one group of producers within the cooperative and particularly failing to fulfill its social responsibility toward those who needed more support in understanding the new production standards. When the program ended, the cooperative was unable to retain the agronomist and accountant and was left without access to credit. The relationship with CEPACOOP thus became of paramount importance for the future of the cooperative.

CEPACOOP was created by the FDC's Technical Assistance Unit to increase the resilience and marketing success of its member cooperatives by providing a stronger voice in price negotiations with Argentinean buyers. CEPACOOP therefore played a crucial role in Guayaibi Unido's development as a cooperative and its ability to transcend local markets. Nonetheless, CEPACOOP's strategy is entirely focused on international marketing, which requires the maintenance of export-quality standards. CEPACOOP left aside broader quality of life issues that are critical for the well-being of cooperative members, including matters related to social equity, solidarity, and the development of management and administrative skills that are essential to the functioning of the cooperative itself. Thus, while CEPACOOP has operated as an intermediary between Guayaibi Unido and Argentinean buyers, it has not provided basic social services and training to cooperative members.

TO HARVEST BANANAS ON A *CHACRA*

To appreciate the basic challenges faced by the cooperative, it is important to understand the process of selecting and packaging bananas destined for the Argentinean market. One of these challenges is the lack of basic harvesting and transportation infrastructure. San Pedro banana producers work *chacras* (small farms) with an average banana crop of three hectares each. Harvesting of bananas occurs throughout the year, although peak production takes place between March and August. The quantity of bananas harvested from each chacra depends, first, on negotiations between CEPACOOP and Argentine buyers. Cooperative managers and the production technician then set per-farm goals based on detailed records of maturation dates, quality requirements, and a fair rotation among producers. Harvesting decisions thus represent a multiscale process that involves all parts of the value chain.

Once the order is given by the cooperative's leaders, the selected producers harvest a determined amount of bananas from their chacras; the next day, workers wash, pack, and load the bananas onto a three-ton truck. A load for export is equivalent to three or four three-ton trucks or 1,050 banana boxes. In this first stage, there are a number of factors that contribute to the final quality of the exported bananas. Poor road infrastructure and excessive handling of bananas can impact quality. Farmers in Guayaibi lack the overhead cable system commonly used to transport bananas across large plantations. High-quality bananas start bruising as they are boxed by hand, then as boxes are moved into the truck, and again as they are driven over several kilometers of dirt roads to the cooperative. Once there, the boxes are once again moved by hand into a larger truck that carries them an additional 10 kilometers on dirt roads before reaching the highway. At the peak of the harvest season it is common to find loaded trucks stuck in the dense clay mud that forms on these roads after a few hours of rain, contributing to further bruising and delays. In addition, production per hectare is comparatively low. As stated by the cooperative's president, "In 2009 Brazil produced 70 tons of bananas per hectare and we produced 22 tons per hectare. Getting 30 tons would be so great, but we have not managed to increase production density." Thus, from the start, the best and "sweetest" bananas, as Guayaibi farmers proudly refer to their product, are doomed. Lack of advanced harvesting and transportation technologies means that these bananas will enter the international market as a second-rate product.

TO EXPORT BANANAS TO ARGENTINA: A RACE AGAINST TIME

Exporting bananas involves a series of bureaucratic processes that essentially constitute a struggle against time in which the quality of bananas and their final presentation is further affected. This is a function of both temperature and the time it takes to transport the cargo. Bananas are harvested prior to reaching maturity and must be transported to a temperature-regulated storeroom where the fruit reaches maturity under controlled conditions. This acclimatization process uniformly transforms a starchy, astringent fruit into the soft, sweet product that consumers look for in the market. From Guayaibi to the Paraguay-Argentina border, the banana must travel 300 kilometers to the town of Alberdi and an additional 1,350 kilometers from the Paraguay-Argentina border to the city of Buenos Aires.

Aside from the considerable distance that must be traveled in trucks that have no climate control, the bureaucracy at the international crossing constitutes an even greater risk since trucks are forced to wait anywhere between 24 and 48 hours at the border. This delay has to do with a two-stage, identical inspection process that occurs on both sides of the border. Before crossing into Argentina, a sample of the content of the truck is unloaded for inspection, and a *precinto* (aluminum band) is affixed to the tarp covering the truck's load to indicate that the merchandise is fine and will not be altered. Once all paperwork is completed, submitted, and approved, the truck is allowed to cross into

Argentina. On the Argentinean side the first precinto is cut, a sample of the merchandise is again unloaded for inspection, and if everything meets expectations, a second precinto is sealed around the tarp. Once all paperwork is completed, submitted, and approved, the truck is given the green light to continue its journey.

The placement of the second precinto significantly complicates the transport of bananas in trucks that lack air-conditioning systems. The driver is forbidden to cut the precinto to open the tarp to regulate the temperature of the load according to weather conditions. For this reason, truck drivers are forced to take the least possible time transporting the bananas. Losing part of the load to overmaturation translates into lower earnings for everyone, including the driver. As one transportation agent put it, "Bananas are a product that won't wait. It is heavily damaged at each [border] inspection and, considering it's our responsibility to transport it in time and proper condition, we are constantly fighting for the border [crossing] to become more consistent." When the merchandise finally arrives at its destination, the importing company cuts the precinto and the banana is promptly transported to air-conditioned chambers for a few hours or days, then distributed to regional markets where they ripen in special chambers until marketed by smaller distributors.

Despite these tremendous marketing difficulties, Guayaibi Unido was able to increase its banana exports from only 66 tons in 2003 to around 3,000 tons in 2007. However, unexpected delays can have devastating consequences. This happened in 2007 when, as described by Vásquez-León (2010), angry Argentinean banana producers, backed by local police and Argentina's federal agency for agro-sanitary enforcement (SENASA), stopped banana imports from Guayaibi at the border, arguing that the bananas were infected with Yellow Sigatoka, a fungal disease. Even though all phytosanitary documents showed that the Guayaibi bananas were clean, the Paraguayan Ministry of Agriculture failed to assist Paraguayan producers in what turned out to be a political move to stop exports to Argentina. Without government support, Guayaibi farmers quickly mobilized a roadblock that lasted five hours and eventually resulted in the opening of the border. However, several tons of bananas were lost, farmers were not compensated, and Argentina ended up introducing additional regulatory inspections. This event contributed to the building of internal solidarity among cooperative members who felt proud of their ability to mobilize, but it also revealed the risk of heavy reliance on MERCOSUR markets.

TO COMPETE IN THE GLOBAL BANANA MARKET: THE ROLE OF PRESENTATION

When the challenges to banana exports to Argentina are cast in a larger context, the pervasive constraints that characterize the reality of small producers in Paraguay (from lack of basic infrastructure, training programs, and access to production loans) become even more salient. Bananas are the most widely marketed fruit in the world with more than 120 producing countries making up the global supply. In the context of MERCOSUR, the small farmers from Guayaibi Unido must compete with highly technified producers from Brazil and Ecuador. Ecuador is the largest producer of bananas in the world with close to 30 percent of global exports controlled by five major multinational corporations, such as Dole and Del Monte. In addition, as one Guayaibi Unido farmer points out, "Banana markets are so unpredictable that when there is too much production from other countries, prices are so low that the wooden packing boxes end up being more costly than the bananas that we put in them." Paraguayan bananas constitute only seven percent of the Argentinean market. In addition to competing with Ecuador and Brazil, Guayaibi Unido producers also compete with the Argentinean banana producers from the northern province of Formosa, the region whose farmers staged the protests described above.

The challenges to maintaining a niche in a highly competitive international market are even greater when quality standards are considered. The producers from Guayaibi Unido proudly characterize their bananas as the sweetest in the Argentinean market. Nonetheless, international standards do not define quality in terms of taste but appearance. In this context, within the Argentinean market, Paraguayan bananas suffer a lower grading because of their appearance. This is a direct result of the conditions of production, harvesting, and transportation in Guayaibi. In contrast, Brazilian and Ecuadorian bananas are produced, harvested, and transported by transnational corporations in plantations with what is internationally considered optimal technology at each step of the chain from production to transportation and final market sale.

Furthermore, despite the constant characterization of Guayaibi Unido as an export cooperative, most of the bananas produced by cooperative members are sold domestically in Asunción's Central Market. Two-thirds of Guayaibi Unido's members market their bananas in Asunción, because so many of them cannot meet the quality standards required for export and must instead sell their bananas locally to middlemen. Despite the actual dependency on national markets, cooperative leaders emphasized that "the objective of the cooperative is to increase production for export and achieve better quality . . . better quality and more production. We will only reach that goal if fifty percent of our member-producers achieve good-quality product on a consistent basis. Right now we are producing eight hundred to one thousand [20-kilogram] boxes per hectare. The minimum that we should be producing is fifteen hundred boxes." Leaders also admit that the educational and solidarity aspects that are fundamental to the workings of the cooperative have been neglected. As one of the leaders stated, "When members see the profitability of international markets they will begin to see the cooperative as an enterprise that belongs to them."

SOCIAL IMAGINARIES OF PRESTIGE

The reigning and pervasive desire to export bananas to the Argentinean market is driven by more than economic incentives. To understand underlying motives, it is necessary to tap into the configuration of local imaginaries—where objects, actions, and people are placed along shared ideals, aspirations, and economic need. Simple material conditions like unimproved dirt roads and the considerable distances between households make motorcycles a necessity for rural households (being by far the most effective mode of transportation in the region). Although motorcycles are a significant economic investment beyond the reach of many families, most people find a way to acquire one, perhaps even diverting part of a production loan. Such commodities as television sets and gas stoves also have great desirability for rural consumers, and the resources acquired to improve the productivity and quality of banana plantations are not always invested toward their original purpose. Moreover, when basic quality standards (set by CEPACOOP) are not met, the cooperative responds by limiting production credit to those producers who have applied the management practices that guarantee quality. Such rationed credit creates friction among those members who feel discriminated against and those who are able to take advantage of the export opportunity.

The production of bananas in the region originally emerged from opportunities perceived in the national market in Asunción, a market without international quality standards. When Guayaibi Unido became part of CEPACOOP, bananas traditionally destined for the national market became characterized as an export product for the MERCOSUR markets. This shift had a strong impact on the social imaginaries of the region. After all, Argentina constitutes the closest representation of modernity in the Paraguayan imaginary. This is a notion that reaches even the most remote regions of the country through the strong presence of Argentinean TV soap operas and the large-scale migration of rural youth into Argentina in search of employment. In effect, to send bananas to Argentina was desirable because it was modern.

The idea of exporting bananas to Argentina (despite the intrinsic challenge of competing in a market completely dominated by Brazil and Ecuador) seduced many producers in Guayaibi in terms of the social prestige implied by such an endeavor. A recurrent theme among cooperative members was "the pride" that resulted from exporting bananas to Argentina. The fact that small-scale banana producers could successfully export their crops despite technological constraints, the lack of basic infrastructure, an overburdened bureaucracy at the international crossing, and competitive market pressures alongside Brazilian and Ecuadorian bananas signified that they were hard-working farmers nobly surviving against all odds. This sense of satisfaction, however, confronts the unyielding reality of increased vulnerability in the international market, where factors are often beyond the members' control.

The most obvious solution to this paradox—product diversification—cannot supplant this push to export, because, I argue, the desire for modernity is too strong a motivation. With such natural advantages such as fertile soils and the availability of water, a place like Guayaibi could produce a wide range of fruits for the national market, but the desire to establish a prestigious international presence and thus portray the face of modernity is too attractive. It is in this context that the challenges faced by Guayaibi Unido as a cooperative are not only external (that is, the challenge of placing their products in a competitive market) but also internal, in terms of social identity and the sense of prestige associated with being an international player. Tackling such issues involves major structural and ideological changes among cooperative members.

CONCLUSION

The success of Guayaibi Unido as an exporting cooperative has been ambiguous. At this point, most cooperative members cannot meet the quality standards of the Argentinean market and the competition of Brazilian and Ecuadorian bananas. The cooperative's product, like that of regional Argentinean growers, is sold as an inferior grade at a lower price. In addition, the specific risks and uncertainties of exporting from Paraguay to Buenos Aires are exceedingly high, increasing the vulnerability of Paraguayan producers. I argue here, however, that the pride and prestige in "exporting bananas to Argentina" constitutes a critical social imaginary for Guayaibi Unido producers and underlies the cooperative's desire to continue expanding into this "modern" market despite the risks.

The actual effects of these imaginaries have greatly impacted the cooperative internally. They have introduced divisiveness within the membership between those producers who have support of the leadership and the means to produce an exportable product and those who do not. In other words, to become part of the exporting impetus of Guayaibi—as desirable as it is for everyone—is not something everyone in the cooperative can aspire to. As a result, two different types of members have emerged: a select few who take part in the exporting activities and those who are not able to access the social, economic, or technological resources to participate in the export market.

The experience of Guayaibi Unido exemplifies the intrinsic challenges faced by smallholder cooperatives seeking to strike a balance between local material realities and local and global expectations. This is a story where adverse material conditions are partially transcended by the will to fulfill larger imaginaries of prestige and modernization. It is sadly ironic, however, that the same banana proudly shipped off by this Paraguayan cooperative is received in Argentina as a lower-grade commodity. How the cooperative is able to overcome these intrinsic technical, bureaucratic, and economic constraints while maintaining a sense of solidarity within the membership will determine the future of both the extant imaginary of modernity and the sustainability of the cooperative.

CAPIIBARY COOPERATIVE

A Strategy of Diversification to Reverse the Rural Exodus

MARCELA VÁSQUEZ-LEÓN

ONE OF THE MOST SIGNIFICANT PROBLEMS affecting the Paraguayan countryside is an ever-increasing process of out-migration. During the last two decades, large numbers of Paraguayan producers have left their agricultural fields in search of labor opportunities in the construction and manufacturing sectors in Argentina and, to a lesser degree, in Paraguay's capital city of Asunción. Confronted with soil degradation and low prices for agricultural products, many farmers see no other option but to sell their land to outside investors seeking to expand industrial soy production. In the past 20 years, the area planted in soybeans in eastern Paraguay has tripled, reaching over 3.5 million hectares (USDA 2015). As a result, this region of Paraguay is progressively losing its small farmers and its forests, as it becomes incorporated into what has been called the "United Soybean Republic" or "Soylandia."*

In the state of Caazapá, where the Capiibary cooperative is located, the problems of out-migration and deforestation due to the rapid expansion of industrial soybean production are severe. Unable to compete, the young are forced

*This refers to the five South American countries (Brazil, Argentina, Paraguay, Uruguay, and Bolivia) that, through the contiguous establishment of industrial soy production as a technologically standardized agro-ecosystem across borders, account for 58.3 percent of global exports (Oliveira and Hecht 2016).

to out-migrate, and the adults left behind are unable to carry on sustained production, leading them to eventually sell family farms. This chapter tells the story of how, over the past 15 years, the Capiibary cooperative has engaged in a multipronged strategy to transform the rural political economy, supporting collective and individual action to increase the profitability and sustainability of small-scale farming and reduce land sales and out-migration among its 1,800 rural members.

A central element of the cooperative's strategy is the diversification of production from a mostly cotton-based farming economy to an economy based on an *estrategia agro-fruti-forestal*, a strategy that combines crop production, fruit orchards, and forestry. In addition, this strategy integrates the household as a unit of production and consumption by paying attention to all phases and types of agricultural production, including those where women have a predominant role. This allows farmers to produce year-round and combines the production of cash and subsistence crops. By extending credit lines along with continuous technical assistance and better access to reliable markets, the cooperative intends to reverse the rural exodus of small-scale producers and address issues of food insecurity and environmental degradation at the regional level. This case study illustrates how cooperative organization and cooperative-led development has the potential to convert small-scale farming into a viable and stable source of livelihood, making farming attractive to young men and women and also enticing migrants to return to the countryside.

COOPERATIVE HISTORY

The Capiibary cooperative was first established in 1988 by a group of local public-sector educators who needed a reliable source of credit at a reasonable interest rate in order to deal with the chronic economic instability caused by inconsistent payments from the state (see parallels with the Coronel Oviedo cooperative, described by Burke in this volume). Thus, in its beginnings, the Cooperativa de Producción Agropecuaria e Industrial Capiibary (Agricultural and Industrial Cooperative Capiibary) was primarily conceived as a credit and savings cooperative to provide assistance to urban members working as public functionaries in the small city of San Juan Nepomuceno, located 250 kilometers southeast of Asunción. Despite its being called an agricultural cooperative, for the first 12 years of its existence cooperative leadership tended to come from the urban sector and rural members received only minor assistance to allow them to produce cotton for a sluggish domestic market.

From the beginning, however, Capiibary had a strong cooperative ethic and dedication to cooperative education. Besides developing into an important savings and credit institution at the local level, in the early 1990s the cooperative founded a school (kindergarten though eighth grade) for children of members and nonmembers, where cooperative principles of mutual aid and solidarity were taught alongside the standard Paraguayan curriculum. The idea behind the school was to expose children to the ethics of cooperativism at an early age and produce a new generation of cooperative leaders. In addition, one of the cooperative founders, a now-retired teacher, also began a cooperative education program, which she continues to offer. Through weekly talks and workshops, she recruits new members and emphasizes the importance of understanding the rights and responsibilities of cooperative membership. While in the past her talks were mostly carried out in the cooperative building, today cooperative educators go to rural areas upon request in order to reach agricultural producers who find it difficult to travel to town. This has been a particularly important strategy to bring rural women into the cooperative and to include them as producers and economic actors. To complement the weekly talks, the cooperative also has a community radio station.

In addition to credit and educational programs, Capiibary provides medical services to its members through a clinic located within the cooperative building, including pediatrics, gynecology, dentistry, and ophthalmology, as well as a clinical laboratory. These services are offered by doctors hired by the cooperative to assist patients once a week. The cooperative also offers a "solidarity service" for those members who may need cash because of an unexpected emergency, such as illness or death in the family.

In 2002, the cooperative changed its name to Cooperativa de Ahorro y Crédito, Producción y Servicios Capiibary (Savings and Credit, Production, and Services Cooperative Capiibary) and began to focus attention and resources on its rural sector, considerably expanding support in the areas of agricultural production, extension, and marketing. Prior to that, cotton was the main cash crop produced not only by cooperative members but also by small farmers throughout Paraguay. Cotton, however, did not provide adequate income nor did it contribute to food security, leading instead to a spiraling cycle of poverty that, at best, allowed farmers to survive when they could count on remittances from migrating children.

Today the cooperative has approximately 6,000 members, which are divided between an urban sector (around 70 percent of members) and a rural sector (about 30 percent). Even though most rural members are small producers with 10 or fewer hectares of land, there are also medium-size producers who own up to 150 hectares. The cooperative's small-scale farmers have reduced the amount of land dedicated to cotton and are instead cultivating citrus fruits as cash crops. Medium producers are combining the production of citrus fruits with grains and forestry. The cooperative's urban members often work as teachers or own private establishments such as retail stores and restaurants in San Juan Nepomuceno, which has a population of around 7,000 people and is considered to be the "economic capital" of the department of Caazapá. The district of San Juan Nepomuceno, however, is mostly rural, with a rural population of about 17,000.

THE CONTEXT OF RURAL INEQUALITY

Caazapá has one of the highest poverty rates in Paraguay. Although the cooperative's small-scale producers benefit from fertile red soils, they must confront the challenges of living along a rapidly expanding agro-industrial frontier for the production of transgenic soybeans primarily by Brazilian farmers who have settled in Paraguay (better known as *Brasiguayos*) and multinational agribusiness. Caazapá, like other rural regions of eastern Paraguay, was negatively impacted by the Paraguayan agrarian policies of the 1970s, which favored the expansion of agro-industrial production of grains for export at the expense of small-scale farming (Galeano 2003).

The soy boom, which began in the 1990s, resulted in the gradual decimation of Paraguay's Atlantic forest and the displacement of thousands of small producers. Since then, Paraguay has become the world's fastest-growing producer of soybeans and the world's fourth-largest soy exporter (Abramson 2009). In 2014–15, Paraguay grew 8.5 million tons of soy per year (USDA 2015) under a model that is incompatible with small-scale farming. Most of the soy is genetically modified and grown for export for animal feed and biofuels. The indiscriminate use of highly toxic agrochemicals has had grave consequences for both environmental and human health. In addition, rapid expansion of soy cultivation has required the cutting of more than 1.2 million hectares of forest. This has led to the dislocation of more than 100,000 small-scale producers who have since been forced to migrate to urban areas as a result of the loss of land and water rights, resulting in a significant decline in the production of foodstuffs for domestic and local markets (Abramson 2009). As Oliveira and Hecht (2016, 252) aptly put it, the soy boom in South America has "converted huge tracts of some the world's most complex ecosystems, some with as many as 600 species per ha, into a monoculture . . . The social life of agriculture and livelihoods in many places has been equally depleted."

This process of out-migration creates greater vulnerabilities for women. Many young women who out-migrate end up as domestic workers in Argentina or Asunción, where they often suffer abuse by employers. They also earn less on average than men for similar work and are the first to be let go during economic crises. Women who have young children are either left by themselves while the male head of household looks for work opportunities elsewhere or are also forced to migrate, leaving younger children in the care of older relatives. With less household income, expenditures on healthcare are usually reduced, further burdening women as caregivers (Floro and Pichetpongsa 2010).

In the rural districts that surround San Juan Nepomuceno, this transformation is evident. Driving across the region, it is striking to see a monotonous sea of soybean fields occasionally broken by tiny islands of native forest. These large fields of hundreds and thousands of hectares are mostly owned by Brasiguayos and a few wealthy Paraguayans. Although Caazapá produces only 6 percent of the country's soy, most of this production is concentrated in and around the district of San Juan Nepomuceno, which borders the largest soy-producing departments in Paraguay. At an average expansion rate of 6 percent per year (USDA 2015), this agro-industrial frontier continues to encroach onto the lands of poor Paraguayan producers and the little remaining native forest. The small agricultural settlements where most cooperative members live are old communities of Guaraní speakers located in the region's rolling hills, where there are larger sections of forest. Some settlements are close to what remains of the largest extensions of Atlantic forest in the area, home of Mby'a and Guayaiki indigenous groups who farm corn, cassava, and yerba mate.

THE COOPERATIVE'S SHIFT

As cooperative leaders began to recognize the economic struggles of small-scale producers and the danger posed by the relentless expansion of the agro-industrial frontier, both in terms of triggering the out-migration of young adults and precipitating the destruction of native forests, the cooperative transformed its vision to incorporate a broad and bold strategy of rural development. During the times that we visited the cooperative between 2006 and 2008, cooperative leaders and members were already implementing the cooperative's new vision of becoming an agent of change in an agrarian context of deep economic inequality by strengthening their focus in the areas of agricultural production, technical assistance, and marketing.

Cooperative leaders began by encouraging producer-members to reduce cotton production, which in 2002 suffered a crisis due to a harsh drought combined with plummeting international prices (Galeano 2003). The cooperative advised farmers to shift to the new agro-fruti-forestal strategy at the urging of the cooperative's dedicated and charismatic agronomist, who spent several years in Brazil learning about what he calls "conservationist agriculture." The cooperative established credit lines exclusively for agricultural producers, including the provision of credit to women's committees, and developed an ongoing program of technical support to help farmers transition to new crops and technologies and eventually increase the quality and quantity of their production. Five years after initiating the agro-fruti-forestal transition, the cooperative's 1,800 rural members were collectively cultivating a total of 1,200 hectares of soy, 700 hectares of maize, 300 hectares of cotton, 80 hectares of passion fruit, and 160 hectares of citrus (grapefruit and orange), in addition to cassava, corn, yerba mate, and other products for household and local consumption. Diversified production started to generate a stable income throughout the year, and in several of the homes we surveyed, young adults had returned to the region to work with their families.

THE ESTRATEGIA AGRO-FRUTI-FORESTAL

The estrategia agro-fruti-forestal, with its three components (citrus production, grain production, and forestry), pairs the ecological principle of crop diversification with the development principle of livelihood diversification, introducing new cash crops in order to reduce ecological stresses and assure year-round income. Through its estrategia agro-fruti-forestal, the cooperative recognizes the importance of developing a diversified model to respond to the various needs of its constituency, since it represents a heterogeneous membership in which land ownership is highly varied in terms of quantity, soil quality, and historical use. After assessing soil conditions, field size, topography, and capital availability, the cooperative presents each member with a custom-made agricultural package. This has allowed the cooperative to serve smallholders and larger producers.

In general, the cooperative encourages and assists small-scale farmers (fewer than 50 hectares) to invest in a diversity of citrus fruits (mainly grapefruit and orange) and passion fruit to allow for year-round income, while also producing subsistence crops like cassava, beans, and corn, and forest products. For example, pairing grapefruit and passion fruit production permits a year-round income and multiyear planning, thereby reducing farmer indebtedness and vulnerability. Passion fruit matures earlier than grapefruits and can be collected twice a year for four to five years. Just as the yield of passion fruit starts to decrease, grapefruit trees reach maturity, providing an income that replaces that from passion fruit until a new cycle begins.

Similarly, the cooperative works to diversify production on medium-size farms (between 50 and 150 hectares) by combining grains, including soy and several varieties of corn,[†] with fruits that have different maturation periods and seasons. The cooperative also has a reforestation program that provides native and nonnative tree seedlings for those producers whose property has an appropriate layout for forest stocks; this provides conservation benefits, wood for household use, and a long-term income source. In addition, through the establishment of public-private alliances and the construction of grain storage infrastructure, the cooperative started to provide marketing options that otherwise would not have been available to the smallest producers of the region.

A key contributor to the growing success of the cooperative is the implementation of a technological package that the cooperative extension agents refer to as conservationist agriculture. This package pursues soil improvement and mechanization as its pillars. To improve production, farmers are trained in the use of green fertilizers and crop rotation, and can access credit for machinery. In order to improve the quality of the soil and prevent erosion, the cooperative's technicians recommend growing green fertilizers (e.g., black oats) together with cash crops. These green fertilizers reduce soil erosion, fix nitrogen, and control the spread of weeds, which reduces farm labor requirements. Green fertilizers are cultivated in experimental plots in the cooperative's green house. This allows the cooperative's technicians to improve their own recommendations through data collected by cooperative extension agents, and gives

---

[†]Although soy is a legume, in Paraguay and many other parts of Latin America it is described as a grain by farmers and consumers.

producers the opportunity to observe firsthand the recommended crops. Although heavy machinery can damage soil, the cooperative's senior technician sees mechanization as a necessary and unavoidable part of this technological package, because it allows for the development and sustenance of small-scale agricultural production, especially for the production of grains. At the time of our visit in 2008 the cooperative was negotiating a new credit line to help producers buy machinery.

A second aspect of the cooperative's strategy is to assist in the storage and sale of grains. Its strategy consists of closely accompanying producers from technical assistance in the production and harvesting phase, all the way to marketing. To improve the terms of commercialization, in 2006 the cooperative built a silo to offer local grain producers the possibility to store their grains until they could command higher prices or producers needed the income. This, together with technical support and broader access to cooperative credit, has allowed Capiibary's small farmers to enter soy production, an enterprise that was thought to be exclusive to wealthy large landholders.

In the midst of heated debates regarding the deleterious environmental and social impact of extensive soy production, Capiibary decided to support and encourage the production of soy by some of its largest members. Under criticism from environmental NGOs, cooperative leaders made a bold decision that has generated positive economic impacts. Cooperative extension agents, however, are aware of the potential environmental problems caused by irresponsible soy production and are working to minimize negative impacts. Through the implementation of its "conservative agriculture" approach, cooperative technicians are working with producers to maintain soil productivity over time and reduce agrochemical inputs by using green fertilizers.

Another fundamental aspect of crop diversification has been the establishment, in 2003, of a public-private alliance between the cooperative and Frutika, a Paraguayan firm that produces fruit juice for sale in Paraguay and Europe. This alliance has been encouraged and supported by the GTZ (German Technical Cooperation) and the departmental government. Frutika was founded in 1977 by a German immigrant family, and its factory, which continues to be run by the family, is located in the neighboring state of Itapúa. Frutika encouraged the production of citrus and passion fruit among small-scale producers by offering a guaranteed market contract and reasonable prices for producers organized in cooperatives. Capiibary was one of the first cooperatives in the region to enter into a formal agreement with Frutika. In the first year, Frutika provided the cooperative with grapefruit and passion fruit seedlings, the required agricultural inputs, and training.

For most cooperative producers, these are new crops. However, an increasing number of farmers have decided to take the risk of trying them out with the understanding that they would be guided along the way by the cooperative's extension agent. Today, the alliance has proven to be a success, benefiting both the producers and the industry. On one hand, small-scale producers have a guaranteed market for their product and a source of steady income for most of the year. On the other hand, Frutika has a constant supply of fruit to maintain its export agreements with the European Union. Frutika perceives the Capiibary cooperative as a responsible organization capable of organizing independent producers and guaranteeing the supply it needs to maintain operations. The downside of this partnership is that Capiibary is relying on only one buyer.

In terms of social organization, the cooperative underscored the importance of establishing local producer committees, especially among those working in fruit production. This type of social structure has proven to strengthen the agricultural sector because it facilitates the delivery of information and technical assistance, and creates an organic system of local collaboration and consultation (see also Burke's chapter on Coronel Oviedo, this volume). Committees have their own structure but are nested within the cooperative, working in close contact with the cooperative's extension agents. By working on committees, producers can also reduce supply, transportation, and marketing costs. For example, producers coordinate their fruit collection schedules and share shipping costs to Frutika.

Emphasis on subsistence production through the support of women's committees, particularly in the small-scale sector, has also been part of the cooperative's new vision. Women have played important leadership roles in the cooperative's history. Several professional women founders, like the woman in charge of education described above, continue to be very active. Similarly, at the time of our visit, the manager of the cooperative was a very dynamic and well-respected female professional. The cooperative has been working closely with women committees to provide credit and training to develop viable projects, like community gardens, chicken and pig production for household consumption and sale, and cassava production for processing into flour. It has also emphasized the creation of microenterprises like the sale of bread, cheese, and jam, as well as support for farmers' markets, where women play an important role.

Finally, the estrategia agro-fruti-forestal also includes the production of trees. Farmers who are interested and able receive native and nonnative tree seedlings from the cooperative. At the time of our visit, nonnative trees like eucalyptus were primarily destined for household consumption, but the cooperative was in the process of venturing into the production of yerba mate, eucalyptus, and native trees for the processing and sale of a variety of forest products. As is the case with soy production, the planting of eucalyptus trees by small farmers has raised eyebrows among environmental NGOs in Paraguay. The cooperative, however, decided to encourage eucalyptus production mainly for consumption as firewood, which is used for cooking in rural households.

LESSONS FROM CAPIIBARY: THE SEARCH FOR SUSTAINABILITY

The Capiibary cooperative, in its search for a sustainable model to empower small-scale producers who face devastation from ever-expanding agro-industry, offers a number of lessons to smallholder cooperatives elsewhere. Through the development of its agro-fruti-forestal strategy, it illustrates how a cooperative can be a learning organization with the capacity to generate and share knowledge in the pursuit of long-term sustainability. Defining sustainable development is difficult and the subject of much debate. Typically, it is defined by a model that balances economic, environmental, and social factors. Sustainability, however, can also be defined by action, by the flexibility, innovation, and adaptation of social and environmental systems over the long term.

As a collective organization, Capiibary has developed a strong cooperative ethic through its multiple cooperative education programs. This ethic comes from the convictions of founding members who are still active in the cooperative and are primarily interested in bringing up a new generation of cooperative leaders with a clear understanding of the principles of mutual aid and solidarity. This is reflected in the cooperative's shift from the urban sector to focus on the agricultural sector, even though the cooperative's leadership has always been and continues to be largely urban. It is also reflected in the importance given to the participation of women as both cooperative leaders and rural producers. The inclusion of women in the cooperative's strategy of rural development not only improves the resilience of women, but also the adaptive capacity of smallholder farming households to all sorts of stressors. In addition, the importance that the cooperative places on supporting young farmers in order to reduce and even reverse the trend of out-migration, as well as supporting rural committees, bodes well for the future of the cooperative's rural sector.

The cooperative's leadership and the charismatic extension agent are bringing innovations in crop types and production systems to the region. In addition, by opening up the possibility of participation in markets from which small farmers have been traditionally excluded (through both the Frutika partnership and soy production), Capiibary can serve as a model for other cooperatives seeking to establish long-term working relationships with national industries in order to find more secure ways to access stable markets. In that sense Capiibary has demonstrated its capacity to share information, respond to interests that go beyond the bottom line, and plan long term with the clear objective of reversing the trend of out-migration.

It is also an organization that is firmly rooted in the district of San Juan Nepomuceno, which means that it understands the predicaments that its members are facing, and as a result, it is more likely to provide viable solutions. It also means that it can be held accountable in terms of its impacts on both the communities and the natural environment on which those communities depend. The cooperative has recognized the importance of developing a diversified model to respond to the various needs of its heterogeneous constituency.

Capiibary also constitutes a rich case study that brings insight into the meaning of grassroots collective sustainable development. Diversifying agriculture among small producers who for decades only produced cotton has required substantial investment in agricultural education and technical support through the implementation of a "best practices" approach, as well as (and equally important) through building social and economic supports to give small farmers a reason to be receptive to these changes. Through implementation of the technological package it calls "conservationist agriculture," the cooperative's development strategy follows important agroecological principles of crop diversification, use of green fertilizers, crop rotation, soil conservation, and reforestation of native species. However, other aspects of Capiibary's rural development model may be perceived as being at odds with the ideology of many peasants' rights and environmental NGOs, who expect small-scale producers to foreground environmental and social responsibility over profitability. At the heart of this disagreement is the cooperative's position vis-à-vis mechanized agriculture, soy cultivation, and nonnative reforestation. As part of its strategy to reduce land sales by small producers and thus contain the expansion of the "soy republic," Capiibary encourages its members to produce soy and also to plant eucalyptus trees. In this context, Capiibary challenges practitioners and academics to question their own often romanticized notions of sustainable development among smallholders.

One of the key challenges that the Capiibary cooperative faces is the ability to go beyond its focus on sustainable economic development and begin a process of political engagement. The expansion of the agro-industrial frontier for the production of soy and the rural exodus of young adults are problems that impact most of Paraguay. The next step for Capiibary would be to begin to actively create a policy environment that supports its goals and strategies for rural development (see also Burke and Piekielek 2011). Engagement in policy discussions about the role and future of the small-scale agricultural sector requires an organization that has robust internal cohesion, a quality that Capiibary has been able to foster and strengthen with its strategy of integrating small producers.

PARAGUAYAN COOPERATIVISM IN CONTEXT

A Practitioner's Perspective

ELISA ECHAGÜE

SMALLHOLDER COOPERATIVES IN PARAGUAY are very heterogeneous organizations, as the different case studies from this section illustrate. In this chapter, I share my particular vision and experience, which I have accumulated over many years working with the smallholder Paraguayan cooperative sector. This experience includes seven years as the director of ACDI/VOCA's Cooperative Development Program in Paraguay coordinating volunteer technical assistance for Paraguayan cooperatives, including those discussed in this volume. My objectives here are, first, to place these cooperatives within the larger institutional context of cooperativism in Paraguay and, second, to highlight the great difficulties facing small-producer cooperatives in general. My intention is to begin a dialogue conducive to the development of support systems that these cooperatives need in order to become independent, long-lasting organizations, with a leadership and membership capable of defining and creating their own development path.

THE INSTITUTIONAL CONTEXT FOR COOPERATIVES

Coronel Oviedo, Manduvirá, Guayaibi Unido, and Capiibary are part of a much larger institutional context that is particularly Paraguayan. The National

Institute of Cooperativism, the supervising organ of cooperatives in Paraguay, estimates that in 2010 the number of cooperatives in Paraguay reached 982, with 1.24 million people associated with the cooperative sector out of a total population of 6.5 million. In other words, around 19 percent of the Paraguayan population belonged to a cooperative association, one of the highest rates of membership in the world. Of the total number of cooperatives in the country, 61 percent are classified as savings and loans, 26 percent are agricultural production cooperatives, and the remaining 13 percent cover a variety of sectors such as labor, medical, and transportation (Carosini 2012). Within these classifications, however, there is a particular character to Paraguayan cooperatives in that the vast majority are multidimensional; that is, a cooperative can simultaneously engage in savings and loan activities, agricultural production, industrial production, and services. Such is the case of the Coronel Oviedo (see Burke this volume) and Capiibary (see Vásquez-León this volume), cooperatives that have a large urban sector and multiple productive activities.

In the case of agricultural production cooperatives, they provide an important contribution to Paraguay's economy, associating approximately 18,000 producers and generating employment for at least 13,000 people (Carosini 2012). In addition, data from the Confederación Paraguaya de Cooperativas (CONPACOOP) indicate that in 2010 agricultural cooperatives generated between 40 to 45 percent of the country's agricultural output and more than 85 percent of milk and milk products, and contributed to 40 percent of the country's exports, 18 percent of Paraguay's GNP, and 26 percent of the agrarian GNP. This is significant given that soybean and beef make up 40 percent of Paraguay's total exports (World Bank 2015). In addition, 24 percent of the credit issued within the country was provided by the cooperative sector, which also housed 16 percent of savings (Colombo and Oxoby 2013).

Production cooperatives in Paraguay are organized into three levels: the first corresponds to individual cooperatives with their parallel areas of activity and membership; the second corresponds to central cooperatives (that group a minimum of three cooperatives) and confederated cooperatives (that group a minimum of seven cooperatives). These, in turn, make up the alliance that represents the third and most encompassing organizational level of cooperatives, CONPACOOP. There are currently seven central cooperatives in Paraguay, which are represented by either savings and loan cooperatives or production cooperatives. Central cooperatives must provide a variety of services to affiliated cooperatives, including the organization of common administration services and the joint use of such services; the provision of appropriate marketing mechanisms and the most advantageous conditions possible for acquiring required goods and services; and the management of support and operations finance services required by associates.

There are three confederations of cooperatives in the country: two savings and loan confederations and one confederation of production cooperatives (FECOPROD). Confederated cooperatives defend affiliates, coordinate actions, and reconcile differences; lend or contract technical assistance as well as support scientific research applied to the activities carried out by the affiliated cooperatives; promote the specialized education of their members; and spread the principles and practices of cooperativism.

Like the agricultural sector in general, Paraguay's agricultural cooperatives are highly heterogeneous and can be broadly divided into (1) those cooperatives with a large membership, agro-industrial infrastructure, and highly sophisticated producers, and (2) those cooperatives that are composed of mainly small-scale, low-technology farmers who belong to relatively small cooperatives, like those discussed in this volume. Both FECOPROD and central production cooperatives tend to be represented by highly developed production cooperatives that receive significant benefits from affiliation. According to the FECOPROD, which groups 33 cooperative organizations, large production cooperatives are responsible for 60 percent of livestock production in Paraguay. These cooperatives count among their members large soy, milk, beef, and pork producers. Their members are primarily Paraguayans of foreign descent, like the Mennonite cooperatives who settled in the Chaco region during the early 1900s and the large cooperatives promoted during the Stroessner government for the marketing of soy. The 2009 annual report of the Inter-American Institute for Cooperation on Agriculture highlights that the production-based cooperative sector in Paraguay is primarily responsible for the dynamism of the agro-industrial sector, as well as that of multinational agro-industrial ventures.

For their part, small-producer cooperatives are located in Paraguay's eastern region. These focus on sesame, fruit, and vegetable production in the northern and central parts of the eastern region and on yerba mate and grains in the eastern and southern parts of the eastern region. The agricultural systems of these producers are characterized by low technology with limited use of mechanization, little or no agricultural inputs, limited credit access, challenges in marketing, and low management skills. CEPACOOP, to which the Guayaibi Unido cooperative belongs (see Rentería-Valencia in this volume), is the only central cooperative that unites a handful of small fruit and vegetable production cooperatives. This central cooperative faces tremendous challenges in supporting affiliated cooperatives.

CHARACTERISTICS OF THE SMALL-PRODUCER COOPERATIVE SECTOR

My role as a technical representative to small producers in Paraguay, especially in the northern region of the country, allowed me to better comprehend the reality and challenges of small-scale producers. The arduous, dogged work done by these men and women deserves to be recognized as an important way of life for many families in the country. Nonetheless, much needs to be done to improve, shore up, and strengthen small-producer agriculture and the standard of living it can provide.

Only a fraction of small farmers belong to a cooperative organization; and even though cooperatives provide important services such as savings and loans, not all of them provide services that specifically target improved agricultural production, processing, and marketing. This makes it difficult to provide an accurate estimate of the proportion of small producers that benefit directly as growers from cooperative membership. Nevertheless cooperative affiliation improves possibilities for accessing markets, getting technical and financial assistance, and advocating for rights and development. There is a need for small producers who are not part of production cooperatives to join forces and begin by forming their own groups or associations.

Still, the expansion of small-producer cooperatives requires an awareness of the variability in cooperative models. Producer cooperatives whose property and management are owned and run by members themselves are fundamentally different from the multiactivity (or multiservice) cooperatives run by professional management staff that include, for example, a livestock and agricultural department. In small-producer cooperatives, usually one-third of the members do business with their cooperative, and in Paraguay, they are called active members. Their capital resources are minimal, for which reason they must turn to external sources, whether banks or public projects, to access production and marketing financing or links to savings and loan cooperatives. The availability of in-house capital is quite reduced, and the attempt to increase institutional capital is challenged by the membership's overall scarcity of resources.

These small-producer cooperatives do not have the financial ability to pay for the services of a professional who might exercise managerial, accounting, and technical assistance duties. Generally, external accountants are contracted for operational bookkeeping while one or more directors manage daily operations. Some of these cooperatives are also linked to the central cooperative that does most of the marketing work. Currently, however, the one central cooperative that represents small producers does not generate sufficient resources to ensure its own sustainability and is highly dependent upon public subsidies.

For their members, small cooperatives provide minimal and irregular services and are often unable to assist all members. Nonetheless, these small cooperatives represent an important alternative for resource-scarce farmers who would otherwise be subject to the discretion of market intermediaries able to manipulate market conditions for their own benefit. Cooperatives constitute a meeting forum for farmers to engage in situational analysis, and as such, they are a valid representational organ for small producers to connect with local, departmental, and central government agencies. Because, however, they are weak and emerging institutions, small cooperatives often lack adequate operational systems and processes and, as a consequence, tend to depend upon external support.

Where leadership in small-producer cooperatives is concerned, the people elected to serve in such posts normally possess a heightened spirit of volunteer service. Although they tend to receive some reimbursements for the costs of their activities in the name of the cooperative, the time they dedicate to such tasks is significant and comes at the expense of personal, family, and work priorities. They are individuals who are recognized in their communities for their social awareness, honesty, academic training, and trustworthiness—all attributes important to effective management. These leaders are clearly committed to the growth of their cooperatives and the improvement of members' quality of life, but for their organizations to grow, they also need a better understanding of basic business concepts and communication skills. In effect, small-cooperative leaders must often play multiple roles—manager, accountant, and secretary—and are thus more consumed by daily tasks than involved in fulfilling the true functions of a cooperative director. This lack of in-house expertise also establishes a significant dependence on external agents contracted by the cooperative.

In most cases, cooperative directors have completed elementary school, and in some exceptional cases, ten or more years of school. They have computing equipment in their offices, but they remain reluctant to learn how to use it from their side of the digital divide. On the other hand, the directors participate actively in the core agricultural and marketing functions of the cooperative and work closely with the membership. Often, they accompany goods to market, negotiate payment, communicate back to members, and in general make key decisions with external agents, such as market intermediaries, input suppliers, warehouse managers, trucking companies, and so forth.

There is a tendency for a limited group of individuals to concentrate power within the cooperative, so that they maintain control over many years. In part, this occurs because most members do not actively seek a director or management position, but where there is a small circle of leaders, there is a risk that the group will become effectively isolated from the larger membership, creating an "us-and-them" environment.

Without a doubt, the membership is the motor that drives the cooperative. It is made up of producers who expect their representatives to do the best possible job directing and negotiating when needed, while they focus on tasks related to the management and care of their land and crops. In general, it can be said that members often display a weakness practicing effective cooperative principles. While a cooperative cannot survive without providing services and benefits to its membership, it is necessary that individual members understand the responsibilities associated with ownership, because they are in fact the owners of the cooperative. In this spirit of ownership, every member must contribute to strengthening the overall cooperative enterprise. However, differences in socioeconomic status can foster tensions when individual members fail to meet their financial obligations or comply with contractual obligations regarding, for example, product quality. In such cases, the entire collectivity is threatened by the actions of individual members. Member loyalty to their organizations is also associated with faithful observance of cooperative principles. In order to market their products at the best offering price, members juggle several different sources of information and connections to their markets. In this context, they can sometimes treat their organizations as one more selling alternative without considering that it is in fact their own business or that their actions determine its success or failure.

In every group, there is a diversity of people with different knowledge and skills. In the Paraguayan context, those people who have little formal education but stand out for the work they do are valued for their *ka'aty* knowledge—a Guarani term that denotes knowledge born of experience. Within the cooperative context, we see that most members do not have a full understanding of financial or business concepts, but some stand out in terms of skills and experience. This allows them to sustain decent living standards that contrast with those of their neighbors.

At the level of production, the system that characterizes small-producer activity in Paraguay is entirely dependent upon nature. The use of irrigation or other types of infrastructure, such as shade cloths, is very limited. Generally speaking, small-producer agriculture is extractive, with minimal return of nutrients to the land. The work that takes place on producer plots is mostly manual labor. In some cases, soil preparation is carried out through mechanization, as is the application of insecticides, especially when it comes to banana production. Harvesting and weed control are done manually.

It could be said that some producers are "true crop specialists." Given that they are in direct contact with their land for hours every day, they know many aspects and characteristics of their crops firsthand, and they carry out corresponding management practices as appropriate. However, very few producers manage their plots according to the established technical criteria for fertilization and prevention and control of infestation and disease. Such external inputs represent costs to the producer and challenges for application. Sometimes, because of economic limitations or by independent decision, producers apply only one of the two or more recommended measures; alternatively, the application of these measures may be ill-timed, or the amount used below what is needed. In these ways, the results are not effective, which in turn generates distrust among producers vis-à-vis the products used. Small producers usually turn to government or project specialists for technical assistance, or specialists from commercial firms provide this support.

ANALYZING THE COOPERATIVE SECTOR: THE COOPERATIVE MODEL FOR SMALL PRODUCERS

The cooperative model was incorporated into Paraguay with the arrival of European immigrants. As such, it is an extremely successful system among producers of foreign background living in the country. Cooperativism is a model that may provide tremendous benefit to the thousands of small producers working in Paraguay, but it is necessary to analyze certain required aspects of its operation so that the system might more effectively serve them. In fact, current experiences indicate that this is possible by connecting more widely developed cooperatives with smaller and emerging ones. Still, if the goal is for a cooperative to be completely independent and self-sufficient, members—whether they are regular members or directors—require training and improvement of business management skills.

Organizational weaknesses among cooperatives and associations of traditional Paraguayan small producers appear in two main areas. First, producers are traditionally individualistic as a result of their lack of access to institutional support. Second, the Paraguayan educational system in rural areas tends to neglect students' self-development by not including organizational, administrative, financial, and responsive training for dealing with the realities of rural production. To launch a cooperative enterprise, one needs not only political will but also the conviction that, through this organizational model, problems affecting small producers can be resolved. In the past, the formation of cooperatives in Paraguay has been promoted and encouraged based on the social benefits that cooperativism generates, sidestepping another very important part of the cooperative experience, which is its business side and the reality that cooperative members are owners, users, and beneficiaries of their own businesses and must generate sufficient income to cover operational and service costs.

Currently, directors and members show heavy dependence on external support in the country. The future of the small cooperatives is threatened by the dearth of strong cooperative education and conviction among members, lack of sufficient financing that would permit autonomous operation, and lack of basic business knowledge and skills. As organizations, producers have the strength of their production, but they lack the organizational and commercial skills required for maintaining a successful enterprise.

Small-scale producers have their dreams but also a limited vision for their organization. Producers should place more value on the strengths and resources available and establish the groundwork for defining a clear vision and mission toward autonomous development of the cooperative. Given their current realities, they have focused mainly on daily tasks and much less on long-term objectives.

The value and importance of cooperative education is fundamental to strengthening production cooperatives. Government programs for promoting cooperative education are currently lacking, particularly in the agricultural sector. Specialists in agricultural outreach are not necessarily experienced in cooperative training. For its part, the cooperative sector itself is implementing at least two support programs for small producers through larger cooperatives who lend them aid and technical assistance. One such program is promoted by the National Institute of Cooperativism and involves assigning support professionals experienced in administration and marketing to small cooperatives. These, in turn, will be linked to cooperatives covering a wider area that will assist growers in working toward ultimate self-sufficiency.

Another area that greatly impacts the economy of small producers is the lack of supportive state policy. It is well-known that smallholder agriculture is highly diversified, but there is a lack of institutional support in the areas of research, infrastructure, and technical assistance that takes the smallholder farm into account. In the areas of marketing and administrative professionals, in Paraguay there is a large number of professional accountants trained to work in commercial agriculture, but the availability of specialists who can work with small producers is very limited. Such professionals must not only use the knowledge that comes from their formal training but also obtain an understanding of the reality of small producers. In other words, they must handle group and organizational dynamics, along with knowing the experiences of rural life. Only by doing so are they able to interact with producers, guided by a heightened spirit of service and effectiveness.

To conclude, I summarize the main challenges facing small producer cooperatives. First, small-producer cooperatives need to be strengthened in order to reduce their heavy dependence on external assistance. Second, cooperative education is fundamental and necessary for these small producers. Third, the business, financial, and administrative skills of directors and members must be enhanced. Fourth, the widespread availability of skilled human resources in cooperative administration and education is crucial to this sector of agricultural production. And fifth, Internet and computing technologies need to be incorporated into the work of cooperative offices. Some of these challenges have been addressed in the Paraguayan case studies described here, but others require further investigation.

As the director of ACDI/VOCA in Paraguay, I had the privilege of working with the directors, members, and staff of these cooperatives. They opened their organizations to us, sharing their problems and needs, showing us their fine human qualities, and participating in that great cooperative spirit that exists in Paraguay. Working with all cooperatives was very satisfying, but it was particularly gratifying to follow such worthy and important efforts like those of the cooperatves that are presented here—the cooperatives of Coronel Oviedo, Manduvirá, Guayaibi Unido, and Capiibary. The Manduvirá cooperative, as explained by Vásquez-León and Gonzalez Aguilera in this volume, recently finished the construction of its own factory in order to *extraer el oro* (extract the gold [from their sugarcane]). In the words of Jackie Theriot, the first volunteer with the Cooperative Development Program assigned to that cooperative, "You all are sitting on a gold mine." These and many other cooperatives are the source of inspiration for envisioning the possibilities of cooperatives as true agents of change.

COOPERATIVE EXPERIENCES IN BRAZIL

COOPERATIVISM IN BRAZIL

The Development Context

TIMOTHY J. FINAN

BRAZIL HAS ALWAYS PRESENTED A DEVELOPMENT ENIGMA. It is the ninth-largest economy in the world (by nominal GDP) and an economic giant in Latin America. The country has almost 9 million square kilometers and a massive endowment of natural resources. Yet fully one third of its nearly 200 million people, and more than half the rural population, live beneath the international poverty line. The major development issue in Brazil has been inequality—both regional inequality between north and south and income inequality between rich and poor. According to the 2006 World Development Report, Brazil is one of the most inequitable societies in the world—in effect, a country that continues to battle widespread poverty in the midst of extensive wealth.

Much of Brazil's poverty is concentrated in the north and northeast regions of the country, where the economies have been based primarily on agriculture, fishing, and extractivism. Although Brazil's demographic structure has shifted from 65 percent rural 30 years ago to 70 percent urban today, industry in the north and northeast has not flourished, and rural areas have experienced little change in agrarian structure or production technology. In sum, poverty and particularly rural poverty are major development challenges in Brazil's northern regions. This poverty contributes to and is exacerbated by massive environmental change of local and global consequence.

In this context of inequality and poverty, the role of cooperativism as a development strategy has both historical and current significance. The principles of cooperativism promote democracy, justice, transparency, and opportunity through association. Cooperativism in Brazil, as elsewhere in the world, emerged as an institution of civil society that would protect the market position of resource-scarce farmers and support small-scale entrepreneurship through force of association. Thus, cooperativism—through its institutional and legal strength—would combat the inequity that left large segments of the rural population marginalized. While the cooperative would be perceived in the marketplace as a capitalist economic entity engaged in production and exchange, it was perceived by the membership as a vehicle for justice and opportunity. This dual role for the cooperative in a development context—successful economic agent and agent of social justice—is what makes cooperativism both unique and challenging in stratified societies like that of Brazil.

COOPERATIVISM IN BRAZIL: AN OVERVIEW

Cooperativism in Brazil is as varied as the country itself. Most historians see the seeds of Brazilian cooperativism in the indigenous communities established by Jesuit missionaries in the early seventeenth century, a model of social collectivism that endured for 150 years. The first cooperative organized under the Rochdale Principles appeared in the state of Paraná in 1847, and agricultural cooperatives were instituted first in the state of Minas Gerais in 1907. Today in Brazil, there are more than 2,000 agricultural and credit cooperatives with over 2 million members. The cooperative movement in Brazil is supported by an established organizational structure at the national and state levels. The Organização das Cooperativas Brasileiras (Organization of Brazilian Cooperatives, or OCB), located in Brasília, represents the official face of cooperativism in the country. It has a strong lobbyist function and advocates for national policies favorable to cooperative interests. Through its private foundation, SESCOOP (Serviço Nacional de Aprendizagem do Cooperativismo, the National Service for Cooperative Learning), OCB sponsors national programs in support of cooperativism, such as cooperative promotion among youth groups. State-level cooperative organizations (Organizaçãoes das Cooperativas Estaduais—OCEs) are part of the OCB network and also support local cooperatives with advocacy and training activities, although state resources tend to be limited. The local cooperatives pay annual fees that help fund these national and state offices. The Ministry of Agriculture in Brasília has a national department of cooperativism (DENACOOP), which has a mission and an annual programmatic budget to foment cooperative development.

Thus, there has been a traditional partnership between government and the cooperative sector. Although the two are officially separate, they are closely interlinked and, in many ways, interdependent. The Brazilian state has accepted the cooperative model as a legitimate mechanism of development, not only in agriculture but in many sectors of economic and social action. The philosophy of cooperativism is publicly endorsed, and cooperative institutions have received significant government support. At the same time, cooperatives are meant to provide the more vulnerable segments of society access to greater economic well-being and to promote social justice in ways that meet national goals. For their part, cooperatives understand that national policy—especially with regard to pricing, investment, and credit—has a vital impact on the success of their organizations.

While this partnership is long-standing, it is also subject to tensions. As is detailed in the four cooperative case studies presented in this section, there are times when government policies have either directly targeted cooperative institutions or indirectly supported their activities through favorable macro and sector-based policies. At other times, national policy decisions (e.g., tightening agricultural credit) have created tremendous barriers to success for cooperatives. Because of their weak institutional development and often precarious financial footing, cooperatives in the north and the northeast have been particularly susceptible to policy impacts. In these regions, it is clear that the government's perception of the role of cooperatives in regional and national development can be a critical variable in determining cooperative success.

The tropical rain forest and *cerrado* regions of Brazil still retain a strong sense of the frontier. Relatively small population clusters tend to be dispersed across a large landscape, and local infrastructure—roads, electricity, and communications—is often inadequate. Periods of intense rainfall impede the movement of goods by land, especially to distant markets. In other periods, searing heat increases the postharvest losses of important crops. Nonindigenous human settlement in these regions is relatively recent, and populations of relatively recent arrivals have often not quite adapted to the unique characteristics of their sometimes fragile environments. Technologies transferred from other regions often do not favor sustainable management of local natural resources. On the socioeconomic side, local institutions are also young and often unstable, and local politics appear more volatile. Educational levels tend to be low, and there is only a small pool of skilled labor. Particularly, management skills are in short supply. Under these formidable contextual constraints, cooperatives are severely challenged to meet their development objectives.

Three of the four case studies presented in this section were initially selected to participate in this comparative study based on several criteria. First, the participating cooperatives were relatively small and were located in areas of high poverty. Also, each one sought to provide their membership an alternative development model. Second, each of the cooperatives had a strategy of economic improvement through greater participation in local, national, and international markets. Third, as stated above, three of the four cooperatives had an existing relationship with ACDI/VOCA, in that they had received technical assistance through the worldwide cooperative development program. The cooperative in Altamira, Pará (see Burke, this volume), was included because of its unique status as an indigenous cooperative integrated into the international cosmetic and health product market. In the following case studies, key determinants of change, including environmental and socioeconomic context, impacts of the broad national policy, dependence on external support, and professional management, are examined within the historical evolution of each cooperative and broader contextual detail. In each, one or another determinant has had a more or less prominent role in the success of the cooperative.

THE EVOLUTION OF COOPERATIVISM IN BRAZIL

JOANA LAURA MARINHO NOGUEIRA AND
JOÃO NICÉDIO ALVES NOGUEIRA

THE BEGINNINGS OF COOPERATIVISM IN BRAZIL are traced to the late nineteenth century in the central-south regions that were mostly inhabited by European migrants with an appreciation for collective action. Sporadic attempts to establish a cooperative institution are documented in the Brazilian states of São Paulo, Minas Gerais, and Rio Grande do Sul, the latter being the site of the first German-inspired credit cooperative, brought to Brazil in 1902 by the Swiss cleric Padre Theodor Amstad and founded under the name Pioneira (Mladenatz 2003). Other historical examples of collective action include a cooperative for telephone workers in São Paulo and a consumer cooperative for public employees in Ouro Preto, Minas Gerais (Panzutti 1999, 2). In fact, however, these evolutionary precursors of the cooperative movement in Brazil lacked two essential characteristics of modern cooperativism. First, there was no body of legislation that had institutionalized collective action and given it a formal identity. Second, early Brazilian cooperatives were not yet imbued with the Rochdale Principles, developed by the Rochdale Pioneers in the 1840s and inspired by Robert Owen, the British social reformer widely considered the "father of cooperativism" (Noronha 1976, 16).

It is important to recognize that cooperativism, as we know it today, began as an alternative to the negative social impacts of the Industrial Revolution, infamous for its wretched working conditions. In Brazil, cooperativism arose first in urban centers but gained its foothold in the rural areas—different from the British experience. From 1906 on, cooperativism became increasingly strong among agricultural producers, many of them immigrants from Japan, Germany, and Italy, where a culture of community action and collective problem-solving was common. These early rural cooperatives embodied the essential elements of the Rochdale philosophy: free and voluntary association, democratic management, economic participation of the membership, autonomy and independence, training and skill building, collective action, and community well-being. These guiding principles gave the cooperative movement its roots and sustained presence in Brazil as in the rest of the world.

INSTITUTIONAL FOUNDATIONS OF COOPERATIVISM IN BRAZIL

The principles of equality and democracy differentiate the cooperative enterprise from private business firms, but these values carry legal challenges in a complex society. According to Perius (2001), Brazilian cooperativism went through three stages: the defining of status within the legal apparatus (1903–38), state interventionist phase (1938–88), and the autonomous self-management phase that has emerged since the Constitution of 1988. The Public Law of 1932 (Decreto No. 22.239) established the unique character of the cooperative as a formal institution in Brazilian society, giving it a status sui generis for all legal purposes (including tax laws), a decision that is still a topic of legal debate in the country. During the 1990s, the government created the institutional machinery wherein the regulation of cooperatives would reside, namely, in the "departments of cooperative assistance" located within the state Secretariats of Agriculture (Pinho 2004).

Beginning in 1938, the cooperativist movement emerged as a subject of state intervention for 50 years, first through the dictatorship of Getúlio Vargas and then the military dictatorship from 1964 to 1985. During this latter period of military control, the state took direct control over the movement as a means of maintaining social order, especially in rural areas, and combating the perceived threats of Cuban revolutionary influence in the poorer rural areas. The General Law for Cooperativism (Lei Geral do Cooperativismo—Lei Nº 5764/71) was decreed in 1971, and the National Council of Cooperativism (CNC) was introduced to set norms and regulations for the sector. The CNC consisted of eight members representing different sectors of government and was presided over by the minister of agriculture. It created a national registry of cooperatives and set all policies and regulations by which cooperatives could operate. With the end of military control in Brazil, a new constitution was ratified in 1988 and remains in place today. Under the constitution, the CNC was disbanded and many of the articles of the Lei Geral were repealed, but the 1971 legislation itself has remained in place.

THE ORGANIZATION OF BRAZILIAN COOPERATIVES (OCB)

Inspired by cooperative leadership in the country, the Organization of Brazilian Cooperatives (Organização das Cooperativas Brasileiras—OCB) emerged in 1969 from the consolidation of two large representative bodies from the cooperative sector. Its mission was, and continues to be, providing technical support to individual cooperatives and the political defense of sector interests at the judicial, legislative, and executive levels of government. According to law, OCB is the formal public representative of the national cooperative system, although its status is that of a civil society entity.*

A major mission of OCB is to lobby Congress in support of legislation favorable to the cooperative sector. In 1986, during the constitutional congress that wrote the new constitution, a parliamentary caucus of senators and representatives supportive of the cooperative movement was formed. Called the Frente Parlamentar do Cooperativismo (Frencoop), it introduced language into the constitution that established the autonomy and independence of Brazilian cooperatives. Under this legislation, any group is free to establish a cooperative without any state authorization and without any state intervention in its

*_in verbis_: "Art. 105. A representação do sistema cooperativista nacional cabe à Organização das Cooperativas Brasileiras—OCB, sociedade civil, com sede na Capital Federal, órgão técnico-consultivo do Governo estruturado nos termos desta Lei, sem finalidade lucrativa, competindo-lhe (. . .) § 1º A Organização das Cooperativas Brasileiras—OCB será constituída de entidades, uma para cada Estado, Território e Distrito Federal, criadas com as mesmas características da organização nacional" (Lei 5764/71, Art. 105). [The representation of the national cooperative system belongs to the Organization of Brazilian Cooperatives—OCB, a civil society nonprofit organization, headquartered in the Federal District and a technical and advisory body to the government according to the terms of this law . . . the OCB network will be constituted of entities located in each state and territory and in the Federal District, created with the same characteristics as the national organization" (Law 5764/71, Art. 105).]

management. Following the ratification of the constitution, the influence of Frencoop was neutralized during a 10-year period but then resuscitated under President Fernando Henrique Cardoso in 1996, and Frencoop now has 238 members, including 215 congressional representatives and 23 senators, representing every state in the country and all major political parties. Frencoop is a major political force in the defense of cooperativism in Brazil and a close ally of OCB.

In contrast to the tenor of cooperative legislation in the United States, which favors flexibility, the Brazilian legislation authorizing cooperative institutions permits only a single organizational model (Bialoskorski 2004, 16–17) and is generally insensitive to differences in type of cooperative and local contextual realities. OCB helped write new legislation, submitted in 2007 as a substitute of the General Law No. 5764/71, but the law has remained deadlocked in Congress. Several pieces of legislation promoted by OCB and supported by Frencoop have been approved in recent years, such as a 2009 law creating the National System of Cooperative Credit. Other legislative efforts have focused on tax laws involving cooperative activity and the division of annual gains.

In accordance with current law, there are 27 state entities divided into five regions that make up the OCB system. They are known as OCEs (Organização de Cooperativas Estaduais—State Cooperative Organizations), and together they account for 7,261 cooperatives with a membership of 8.2 million and almost 249,000 employees (see table 8.1). The cooperatives pay annual dues to support the OCB.

These numbers reveal a clear pattern of regional variability in the influence of cooperativism accross the country. The south accounts for 18 percent of all the cooperatives and 38 percent of all the membership in Brazil, which reflects the historical influx of post–World War II farmers from Europe bringing with them a tradition of collective action. The southeast of Brazil, including the states of São Paulo, Rio de Janeiro, Espírito Santo, and Minas Gerais, accounts for 37 percent of all cooperatives and 50 percent of membership. In part, the predominance of cooperativism in this region has historical roots in the immigration of Japanese workers bound for the coffee plantations in São Paulo. From the 1890s to the late 1930s, over 600,000 Japanese landed in Brazil, and the majority soon abandoned the harsh life in the coffee fields to occupy the then-frontier regions where they became independent farmers. Perhaps due to their cultural isolation, these farmers routinely formed agricultural cooperatives, some of which became among the largest cooperatives in Brazil.

TABLE 8.1. Numbers of Cooperatives, Membership, and Employees by Region and State (2009)

REGION	STATE	N	MEMBERS	EMPLOYEES
SOUTH	Rio Grande do Sul (OCERGS)	799	1,738,510	45,874
	Santa Catarina (OCESC)	257	858,671	29,924
	Paraná (OCEPAR)	238	513,771	55,367
Subtotal		1294	3,110,952	131,165
SOUTHEAST	São Paulo (OCESP)	905	2,822,202	59,126
	Rio de Janeiro (OCB-RJ)	815	268,235	8,222
	Minas Gerais (OCE-MG)	789	902,749	26,239
	Espírito Santo (OCB-ES)	143	147,127	5,641
Subtotal		2,652	4,140,313	72,989
CENTER-WEST	Distrito Federal (OCDF)	154	142,180	1,774
	Goiás (OCB-GO)	225	82,020	6,043
	Mato Grosso (OCB-MT)	159	183,957	6,139
	Mato Grosso do Sul (OCB-MS)	105	77,752	3,322
Subtotal		720	495,329	17,926
NORTHEAST	Alagoas (OCB-AL)	98	19,986	3,341
	Bahia (OCE-B)	820	73,229	2,668
	Ceará (OCE-CE)	154	67,243	5,582

continued

TABLE 8.1. *(continued)*

REGION	STATE	N	MEMBERS	EMPLOYEES
	Maranhão (OCE-MA)	244	12,636	676
	Paraíba (OCB-PB)	115	45,768	2,581
	Pernambuco (OCB-PE)	199	105,268	3,126
	Piauí (OCB-PI)	75	15,243	686
	Rio Grande do Norte (OCB-RN)	124	58,169	1,297
	Sergipe (OCESE)	60	10,913	585
Subtotal		1,889	408,455	20,542
NORTH	Acre (OCB-AC)	48	7,385	195
	Amapá (OCB-AP)	47	4,622	450
	Amazonas (OCB-AM)	165	12,047	1,485
	Pará (OCB-PA)	264	45,566	1,669
	Rondônia (OCB-RO)	124	24,208	1,182
	Roraima (OCB-RR)	58	3,533	348
	Tocantins (OCB-TO)	77	9,420	648
Subtotal		783	106,781	5,977
Totals	Brazil	7,261	8,252,410	247,951

SOURCE OCB, 2009

The nine states of the northeast region, on the other hand, did not experience the contribution of European or Japanese immigrants. This region has a large number of cooperatives (25 percent of the total) as the result of a state-led effort to expand cooperatives in the 1970s in response to a perceived Cuban threat, but cooperative membership is only around 5 percent. The poor acceptance of cooperativism in the region is partly due to its imposition by the state, the lack of a tradition of collective action, and underlying socioeconomic conditions. The northeast is the poorest region of the country and is characterized by a semiarid environment subject to severe and frequent droughts. Until the last 40 years, the majority of the population lived in rural areas and made their living from rain-fed agriculture and livestock ranching, even though the region offers limited conditions for a vibrant agricultural economy. One of the case studies in this volume (Barros) focuses on a cooperative in the northeast state of Ceará.

The seven states of the north, whose combined territory includes the majority of Brazil's Amazon rain forest, also have a weak cooperative tradition. Together they account for 10 percent of the cooperatives but only one percent of total membership. Nonetheless, the underlying potential of the region's resource base provides a motive for successful collective action. The region has the opportunity to lead in sustainable development strategies for fragile ecosystems, and cooperatives have a role to play in this process. In this volume, three case studies focus on cooperatives from the Amazonian north.

IMPACTS OF COOPERATIVISM

Cooperatives have a presence in 1,700 *municípios* of Brazil, around one third of the total.† Although agricultural cooperatives have dominated the sector historically, Brazil classifies 13 types of cooperatives by economic activity, including credit cooperatives (3.5 million members), consumer cooperatives (2.3 million), infrastructure cooperatives (716,000), labor cooperatives (261,000), health cooperatives (226,000), housing cooperatives (110,000), and so forth. There are 1,615 agricultural cooperatives, with 942,000 members and 138,000 employees.

†In the Brazilian politico-administrative system, the município is the lowest level of administration and is led by an elected official (a *prefeito*) and a municipal council (*câmara dos vereadores*). It is similar to a county in the United States.

TABLE 8.2. The Município-Level Impact of Cooperative Presence on the Human Development Index Scores

REGION	CENTER-WEST	NORTHEAST	NORTH	SOUTHEAST	SOUTH	BRAZIL
Number of Cooperatives	643	1,889	783	2,652	1,294	7,261
% Municípios with Cooperatives	29.4	31.0	36.3	32.2	31.3	31.7
HDI (Municípios Without Cooperatives)	0.73	0.60	0.65	0.73	0.76	0.67
HDI (Municípios with Cooperatives)	0.76	0.63	0.70	0.76	0.80	0.70
Difference in HDI	0.03	0.03	0.05	0.03	0.04	0.03

SOURCE UNDP, the Brazilian Statistics Agency (IBGE) and OCB.

In recent years, however, there has been a resurgence of urban labor and credit cooperatives. Overall, in 2009, the cooperative sector generated $44 billion in economic activity, including $3.5 billion in exports, primarily of the agricultural commodities sugar, ethanol, soy beans, and coffee. A study conducted by the United Nations Development Program suggests that cooperatives bring benefits at the local, município level. Using the Human Development Index as an indicator, those municípios with cooperatives scored consistently higher by about 4 percent (see table 8.2). It is possible to conclude that these differences are due to the higher salary levels, the valued-added of local agricultural commodities, the regular supply of inputs and services, and the lower interest rates from cooperative financing. In effect, the flow of resources to cooperatives brings benefits both to the membership and to the local community through various multiplier effects.

THE SERVIÇO NACIONAL DE APRENDIZAGEM DO COOPERATIVISMO (SESCOOP)

SESCOOP is an agency within the national cooperative movement created in 1998 under Provisional Law 1.715 and regulated under Decree Nº 3.017, which provides broad professional and technical support to cooperatives and leads the national effort in cooperative awareness and education. SESCOOP has local offices in 27 states and in the Federal District, with a central office in Brasília. If OCB is more the political arm of cooperativism in Brazil, SESCOOP provides a wide range of services to local cooperatives seeking to expand the reach of cooperative action through education, social assistance, and technical assistance to member cooperatives. In this sense, SESCOOP fills an important gap by offering training to managers and members, cooperative education, and direct monitoring of local cooperative progress.

It is characteristic of Brazilian cooperatives that a member of the governing board (an elected position) is also the cooperative manager. While in U.S. and European cooperatives the management of the organization is the responsibility of trained professional managers who carry out the policy decisions of the board, Brazilian cooperative managers are usually not professionals trained in business and cooperative principles. Thus, the main challenge of SESCOOP is to provide the needed professional training and to encourage the separation of decision making and management.

GLOBAL PARTNERSHIPS IN BRAZILIAN COOPERATIVISM

OCB and SESCOOP also promote partnerships that bring member cooperatives into wider international networks. One example has been the lasting partnership with the nonprofit organization ACDI/VOCA, which uses volunteer technical advisors to share the experience of North American cooperatives with local counterparts in Brazil, especially those in the north and the northeast. This relationship began in 1973 based on a request for assistance from the Secretariat of Agriculture of the state of Minas Gerais. At the same time, a program of cooperative assistance was created in collaboration with the state extension services, and OCB covered the office costs of ACDI/VOCA during the period between 1995 and 2007. Under this partnership, 186 technical assistance missions were carried out from 1973 to 2007, most of them in the north and northeast regions.

As part of its global strategy, OCB is an active participant in OCPLP (the Organization of Cooperatives for People of the Portuguese Language) and RECM (the Special Conference of Cooperatives of the MERCOSUL). The OCPLP is located in Lisbon and joins the eight Portuguese-speaking countries in Africa, Latin America, Asia, and Europe in a common effort to support cooperativism. For its part, RECM is part of the MERCOSUL common economic zone and consists of cooperative organizations from Brazil, Paraguay, Uruguay, and Argentina as permanent members, while Venezuela is pursing membership and Chile and Bolivia are formal observers. Since 1988, OCB is also a member of the International Cooperative Alliance (ICA), the most important international representative of the cooperative movement. Roberto Rodrigues, president of the OCB from 1985 to 1991, is the only non-European leader to ever head the ICA (from 1999 to 2001). The current OCB president is also the vice president of the regional ICA for the Americas.

CONCLUSION

The Brazilian cooperative movement has gone through a dynamic process of development, expansion, and adjustment to the broader realities of Brazilian society and economy. Much of this process has involved creating independence from the state while using state resources to enhance professionalism of the sector, not only in terms of cooperative management but also in terms of commercial activities. The challenge is to create a sustainable and viable economic enterprise that can compete in a modernized Brazil. Cooperativism has also accompanied the demographic transformation of Brazil into an urban society, and the number of urban-based cooperatives has risen rapidly. As it adjusts to the flow of history, Brazilian cooperativism now occupies an increasingly important role in the development process. Its challenge is to lead the drive toward economic, social, and environmental sustainability into the twenty-first century while upholding and enhancing the principles that set cooperativism apart.

ATOTORI

The Discourse and Practice of Sustainability in the Amazon Rain Forest

JESSICA PIEKIELEK AND TIMOTHY J. FINAN

ONE OF THE BEST-KNOWN AGRICULTURAL COOPERATIVES in Latin America is CAMTA (Cooperativa Mista Agrícola de Tomé-Açu, or the Mixed Agricultural Cooperative of Tomé-Açu), located in the *município* of Tomé-Açu in the state of Pará, Brazil. Its reputation derives from its unique origins, its success in the face of multiple obstacles—both economic and political—and its vision of discovering a sustainable agricultural model for a fragile tropical environment. The story of CAMTA has been retold multiple times in print and film, in both academia and in global commerce. In 1929, a few dozen Japanese immigrant families eschewed the relative comforts of a new home in São Paulo (in the south) and departed on the adventure of settling virgin rain forest along a river about 260 kilometers from the state capital of Belém in the north of Brazil. Through periods of boom and bust, this *colônia* of Japanese farmers created a cooperative built on a base of social capital and identity, and today CAMTA is the Amazon's largest exporter of tropical fruit pulp.

In contrast to failed attempts to exploit the fabled forest wealth (e.g., Jarilândia and Fordlândia),* CAMTA has created a sustainable agroforestry

---

*For Jarilândia, see Jordan and Russell (1989); for Fordlândia, see Grandin (2010). Both these undertakings were massive in scale and now lie as witness to the hubris of humankind.

model that mimics the diversity of the ecosystem while producing economic benefits to farmers. As the managing director of CAMTA has stated in interviews, the history of CAMTA is interwoven with the history of the Japanese immigrant members. This chapter explores that history with a focus on how essential Japanese values forged a shared vision of sustainability and continuity among the CAMTA membership, thus strengthening the cooperative as a local economic institution. As the cooperative prospered, it increased the range of noneconomic services provided to its membership, including health, education, access to commerce, and cultural events, thus reinforcing the close linkage of local identity, welfare, and the success of the cooperative. At the same time, this chapter derives from the CAMTA story a lesson in adaptiveness and resilience in the face of internal and external stresses. This chapter is based in part on field research, including interviews, member livelihood surveys, and observations, conducted by the authors in the summers of 2005 and 2009. Methods and results from a survey of cooperative members' livelihoods are reported in Piekielek (2010).

CAMTA: ORIGINS

It is difficult to imagine the range of emotions of those first 43 immigrant families when they disembarked at the edge of the Acará-Mirim River to open up virgin rain forest that had been negotiated for them by the Companhia Nipônica de Plantações do Brasil (Japanese Plantation Company).[†] Each family received a plot of 25 hectares for which annual payments were made (Piekielek 2010; Tafner and Silva 2010). At the urging of the company, the first immigrants planted cocoa as a cash crop and rice and vegetables for home consumption. In 1931, the colony formed a *cooperativa de hortaliças* (vegetable cooperative) to market produce in the city of Belém, 12 to 18 hours away by boat. The consuming population of Belém had neither refrigeration nor a tradition of vegetables in their diet, thus this venture encountered many difficulties.

[†]In fact, the company, known as Nantaku—its Japanese acronym—was a private firm subsidized by the Japanese government but motivated by the potential profits from colonization. Nantaku provided technical and marketing assistance to the early settlers but always with an eye toward reward (Tafner and Silva 2010).

The gargantuan task of clearing the rain forest and the low return to this labor were further exacerbated by the incidence of malaria, which was widespread and caused many deaths among the early settlers. Subsequent ships arrived with their immigrants, but the difficulties of life in the rain forest began to dissipate the initial hope and enthusiasm, and many immigrants ended up leaving the region for São Paulo or elsewhere (Tafner and Silva 2010, 129). Nonetheless, between 1929 and 1937, 21 ships arrived at the settlement carrying a total of 361 families.

In the context of the worsening economic situation, tensions arose between the company and the early settlers, particularly regarding the 30 percent of all production that the company exacted. The company abandoned Tomé-Açu, and the settlers assumed all responsibility (and decision making) for their well-being. With the fall of the formal settlement system, there was widespread abandonment of the region, and only 98 Japanese families remained in Tomé-Açu in 1942.[‡] A key binding factor that formed the core of this diminished community was the cooperative. A vegetable market had been gradually created in Belém, and the cooperative members began to expand the diversity of marketed crops to the point that the vegetable cooperative was renamed the Cooperativa Agrícola do Acará (Agricultural Cooperative of Acará). Increasingly, according to interviews with CAMTA leaders, the cooperative and the community became indistinguishable.[§]

FROM BLACK GOLD TO *AÇAÍ*

In 1942, Brazil entered World War II on the side of the Allies, and all Japanese immigrants in the Amazon were confined to Tomé-Açu, which was converted into an internment camp. The Cooperativa do Acará was taken over by the Brazilian government and turned into the administrative arm overseeing the camp residents. In effect, the cooperative was made a state monopoly (called CETA,

[‡]During this time period, 288 families, accounting for around 1,600 people, left the region (Tafner and Silva 2010, 133).

[§]In interviews in 2005, older cooperative members often did not distinguish between the cooperative and other community organizations, attributing to the cooperative a range of benefits and projects that were sponsored by other Japanese immigrant community organizations.

Companhia Estadual de Tomé-Açu) with total control of local economic activity. At the end of the war, the camp was dissolved and the immigrants pressured the state to return their cooperative. In a show of great solidarity, the Japanese community built its own motorized boat to resume its marketing activities to Belém. This demonstration of independence reinforced the value of collective action among the remaining families of this tight-knit community.

In 1933, the fourth ship carrying immigrants to the Amazon stopped in Singapore to bury a passenger who had died, and one of the immigrants, Makinossuke Ussui, acquired a small number of black pepper plants (*Piper nigrum L.*) and planted them in an experimental field station run by the cooperative. At the end of the war, black pepper was increasing in value and importance, and the Japanese farmers of Tomé-Açu were incorporating more of this cultigen into their fields. The community, bolstered by signs of economic improvement, met regularly to discuss the creation of a new cooperative, but this time a cooperative that was officially recognized by the Brazilian government and complied with existing cooperative legislation. In 1949, CAMTA was officially founded under the management of young community leaders who brought both a sense of history and tradition and a vision for a future of greater inclusion in the national and global context. Since its inception, the cooperative has been run democratically, on a one member, one vote model, with management elected from the member base.

According to Homma (2004), the decades of the 1950s and '60s created an economic windfall for the CAMTA community. With the wartime destruction of pepper vines in Southeast Asia, the price of black/white pepper grew exponentially, from $220 per metric ton in 1945 to over $3,000 per metric ton in 1950 (Tafner and Silva 2010, 140). One issei elder from the community told us in 2009 that his large house had been built with pepper revenues and that, during this boom period, each ton of pepper purchased a truck. Tomé-Açu became the largest exporting region of black pepper in Latin America and provided up to 5 percent of the global supply. CAMTA exported the pepper of its membership, and the cooperative increased its economic prowess in the município of Tomé-Açu, whose population rapidly grew with the arrival of non-Japanese Brazilians in search of work and economic opportunity.

The confluence of CAMTA and community was manifest during these "golden years" (roughly 1950 to 1970) in the large investment in quality of life for the membership. CAMTA built a hospital, a malaria eradication clinic, schools, and a cultural association in which community solidarity based on shared

Japanese immigrant identity was reinforced and celebrated. The cooperative also opened and managed businesses to support members, such as a general store, agricultural supply store, and gas station. Later, a golf course and baseball stadium were built in the rain forest for leisure pursuits that are uniquely Japanese. In this sense, the cooperative penetrated multiple aspects of everyday life, and the activities of CAMTA and those of the immigrant community became virtually seamless. Of equal importance, these projects benefited a range of members across income brackets.

The pepper boom ushered in a period of technological change among the producing community. Farmers adopted the widespread use of chemical fertilizers and other inputs, and many moved toward a pepper monoculture. In the 1970s, a fungus, *Fusarium solani*, began to devastate large plantations of pepper vines, and with large fluctuations in the global pepper price and increasingly rampant inflation (affecting currency exchange rates), pepper producers began to experience financial crisis. CAMTA had received regular technical and financial assistance from the Japanese Emigration Services, which later became part of JICA (the Japanese International Cooperation Agency). With technical support from JICA, the cooperative sought to devise alternatives to a pepper monoculture, and many CAMTA members adopted cocoa (*Theobroma cacao*) into their production systems (Saes et al. 2014).

A further push toward diversification occurred with the commercial adoption of *cupuaçu* (*Theobroma grandiflorum*), a tropical relative of the cocoa family, introduced by CAMTA pioneer Katsutoshi Watanabe in the 1980s (Homma et al. 2014). At the same time, with financing from JICA, a fruit pulp plant was established under the management of a new cooperative called ASFATA (Associação de Fomento de Tomé-Açu, or the Tomé-Açu Development Association), with this membership being nearly identical with that of CAMTA (Piekielek 2010). The fruit pulp factory began to supply a wide range of fruit juice products to the Brazilian domestic market. In 1991, CAMTA purchased the fruit processing plant, disbanded ASFATA, and borrowed funds from JICA to expand the cold storage capacity. CAMTA producers began to produce a wide range of fruits, such as melon (*Cucumis melo*), papaya (*Carica papaya*), passion fruit (*Passiflora edulis*), pineapple, guava, and the "tropical cherry," acerola (*Malpighia glabra* L.).

Despite CAMTA's promotion of diversification, members still relied heavily on black pepper, the price of which fluctuated widely. As Brazil entered a period of hyperinflation and severe credit constriction in the 1990s, the cooperative experienced a major crisis. It was laden with debt, and the value of its stocks was declining due to low global prices and domestic inflation. The crisis precipitated a change in cooperative management and an emergency meeting of the new directorship to determine if CAMTA should disband.** The new CAMTA leaders decided to sell their stored inventory, to divest many of their holdings (the supermarket, gas station, supply store, outlets in São Paulo, etc.) and to ask the membership to come up with funds, in cash, to cover the rest of the debt. In a show of solidarity and trust, the membership accepted this decision, and the cooperative continued its operations under a new, austere set of conditions.

This time, however, shifts in the economic environment brought luck to CAMTA. In 1987, JICA had provided financing for rural electrification. Now this credit was used to construct the infrastructure necessary to connect cooperative members to the grid, which was supplied by the public state energy company. A cooperative, COERTA (Cooperativa de Electrificação Rural de Tomé-Açu, or the Tomé-Açu Rural Electrification Cooperative),†† managed the electrification project, and then sold the infrastructure to the state company in exchange for shares that were virtually valueless. Then, in the 1990s, the national neoliberal trend in Brazil was to privatize much of the economy, including the energy sector. The Pará state energy company put its stock on the market, but only to those who owned shares in the company. COERTA, as a shareholder, saw the value of its stock explode upward and auctioned shares off for more than $3 million. This capital was made available to CAMTA, which used the money to construct and equip a new fruit processing plant with a production capacity of 5,000 tons of fruit products (Piekielek 2010, 19). In effect, CAMTA institutionalized its commitment to diversification at both the production and marketing levels.

This shift toward a diversified product range is embodied in CAMTA's response to the rapidly growing market for açai (*Euterpe oleracea*), a globally popular palm fruit native to the Amazon and known for its health-enhancing characteristics. Açai trees produce twice a year but only starting in the third year of growth, and optimal harvests are not achieved until the sixth year; thus, it is a longer-term investment on the part of farmers. Native stands of açai are

**It is important to note that at this time, several large Japanese cooperatives failed and were shut down, including the iconic Cotia cooperative in São Paulo.

††Again, the membership of COERTA was mostly identical with that of CAMTA.

found throughout Amazonia where soil and moisture conditions are favorable. The small purple drupe (fruit) forms in clusters at the top of the palm, and their harvest by hand requires climbing the trunk. Of critical importance, the fruit requires processing within 24 hours of harvest, more easily facilitated by the cooperative's local fruit-processing plant. CAMTA has acquired an organic seal for its açai and has become a recognized national and international supplier. In 2007, açai accounted for more than half (57 percent) of the cooperative's total fruit pulp sales, with most being exported to the United States and Japan (Tafner Jr. and Silva 2010, 149). More recently, the cooperative partnered with a chocolate producer in Japan to produce fine-quality cocoa beans. After making improvements in fermentation, CAMTA received the International Cocoa Award in 2010, and a quarter of the cooperative's cocoa beans went to produce premium chocolate in Japan (Saes et al 2014). The açai and cocoa stories exemplify CAMTA's ability to reconcile production challenges with expanding markets.

SUSTAINABILITY IN A RAIN FOREST

Sustainability is a complex concept that can assume various dimensions. It implies a record of success through time without reduction in the means of that success. Thus, there are economic, environmental, and quality-of-life components that make a social-economic-ecological system sustainable. In this more complex sense of the concept, the role of CAMTA in discovering a sustainable coexistence with an essentially fragile ecosystem is a core message of this chapter. We propose that sustainability should be assessed from a long-term perspective of institutional, environmental, and cultural continuity.

Institutional continuity is closely related to economic success. In 2009, CAMTA had approximately 135 members (up from 113 in 2005), 75 percent of whom were of Japanese descent. Its diverse operation maintains a receiving warehouse and marketing infrastructure for black pepper, cocoa, and cupuaçu. The black pepper is cleaned, classified, and sacked for export, while the cocoa and cupuaçu yield multiple commercial products, including pulp, beans, and butter from crushed beans. CAMTA is also considering the production of a chocolate made from cupuaçu (called "*cupu-late*" in Portuguese). The fruit pulp processing plant has 14 fruit products and in 2009 produced 3,500 tons of fruit pulp, about 70 percent of capacity. In 2007, the sales of fruit pulp represented around two-thirds of the total revenue flow of the cooperative.

The constraining factor to achieving full capacity in the pulp operation is the availability of quality raw material. The rural population of Tomé-Açu and neighboring municípios has expanded with the economic growth associated with CAMTA; however, in contrast to the relatively wealthy and technologically sophisticated CAMTA membership, most regional farmers practice a small-scale, low-input agriculture and are resource-scarce. Whereas the CAMTA membership has not been able to fulfill the demand for the range and quantity of fruits processed in the plant, the small-farm sector does not possess the resources and the technology necessary to provide a steady and reliable flow of quality raw material (Homma et al. 2014). Thus, the institutional challenge is to engage an ambitious strategy of technology transfer to integrate potential producers into the stream of products to the cooperative.

The CAMTA management is well aware of the historical struggles to achieve a viable institutional presence, and the role of technology adoption has always been recognized and valued. The cooperative has provided a research and development capacity from the time that the first pepper vines arrived from Singapore. This openness to new knowledge from whatever the source has been a key mark of the CAMTA approach. Since the establishment of the processing plant, the cooperative has created a rural extension sector that works with associations of small producers in the neighboring regions, and CAMTA members provide regular technical assistance to these associations. Interviews in 2009 with one specific association, the Associação da Quarta Região (Fourth Region Association), revealed that CAMTA had played a major technical and organizational role in transforming the production capacity of the group. CAMTA management's strategy is to invite the associations to become institutional members of the cooperative once a certain level of technical expertise has been achieved. It appears clear that the institutional sustainability and economic growth of the cooperative will depend upon the ability to incorporate a wide swath of producers into its supply catchment area, and that this process will in turn spread the benefits of the cooperative.

The second area of sustainability is environmental. Early in the history of the cooperative, one of its prominent members, Noboru Sakaguchi, who had studied agriculture in Tokyo, was inspired by the boat trip from Tomé-Açu along the Acará River through the rain forest to Belém. He noticed the riverbank farmsteads with many different species of trees, bushes, and plants all mixed in the same garden (Yamada 1999). He observed that there were always some species in production throughout the year and wondered if this diversification strategy could serve as a wider model of land use in the Amazon rain forest. Sakaguchi is credited as one of the pioneers of what is call the "agroforestry system" (SAF—Sistema Agroflorestal), a land-use system developed by cooperative members and now a focus of intense global interest (Yamada and Osaqui 2006; Yamada and Gholtz 2002a, 2002b; Porro et al. 2012).

The SAF, in contrast to the black pepper monoculture, features a crop mix of long-growing (and prized) timber species, many of which were cleared in the earlier occupation of the rain forest, with annual producers (once mature) açai, cocoa, and cupuaçu, along with papaya, acerola, passion fruit, banana, melon, and other fruits. The planting and growth sequence mimics that of rain forest progression from herbaceous communities to arboreal ones (Bolfe and Batistella 2011). We interviewed farmers who planted pepper vines and passion fruit in a cleared field, then interplanted species such as cocoa and cupuaçu between the rows, then interspersed the "noble" timber varieties (mahogany, *ipé*, *andiroba*, *cedro*, etc.) and Brazil nut throughout the field. This system produces an income stream in two to three years and continues producing up to 30 or more years when the timber species are ready for harvest.

There is great variability within the SAFs themselves, and Bolfe and Batistella (2001) have characterized four SAFs in terms of their cover structure and productivity. Yamada's 1999 study describes the diversity of the SAFs in Tomé-Açu as follows:

> The crop species comprising these systems included three perennial vine species, four species of shade trees, 33 fruit tree species, 68 MPT species and numerous vegetable, herb, grain, tuber and green manure species. Over 300 different polycultural combinations of 70 species, 90 percent of which are trees, were documented in the main production fields (excluding home gardens) of Tomé-Açu farmers. (Cited in Yamada and Gholz 2002, 18)

There is widespread consensus that these SAFs are environmentally sustainable, and, in fact, the CAMTA model has been disseminated around the world in person, print, and film. It is important, however, to acknowledge that the SAF is not a "natural" system but one designed and managed by humans. In this case, we ask what incentives exist for a middle-aged farmer to plant a species, such as mahogany or even Brazil nut, that yields financial returns only after 30 years. In this regard, the underlying values attached to sustainability become the focus of analysis. In interviews with CAMTA members in 2009, the Japanese concept of *atotori* arose repeatedly. In Japanese family structure, the atotori is the successor, the one that carries on a family tradition and protects and cares for the family legacy (Kitaoji 1971).

In Tomé-Açu during the 1990s, many *nissei* young adults left Brazil to take manual jobs in Japan in order to accumulate capital.[‡‡] In 1993, Paulo Gustavo reported that, of 1,300 Japanese immigrant descendants from Tomé-Açu, 400 were in Japan. Despite the long-standing formal and informal ties with Japan, the return migration has been mostly temporary. There is great commitment to maintaining the legacies created by fathers and grandfathers. Children know that they will benefit from the timber investment made by their fathers—plantations of mahogany, Brazil nuts, and other slow-growth timbers—and they, in turn, will plant more for their children. During interviews, several older members spontaneously shared the importance of passing on their knowledge to future generations, especially through the cooperative and its youth programs. CAMTA has a very active youth association and supports a wide range of mixed cultural activities, from baseball to São João celebrations,[§§] all reinforcing a sense of community and the importance of atotori. In an interview, one member of the youth group stated, "We have to honor what our father has done for us, what he has established." Thus, environmental and institutional sustainability is reinforced in a cultural system that provides deep-seated and essential incentives to protect and enhance the livelihoods established by previous generations.[***]

[‡‡] These return migrants of Japanese descent are called *dekassegui*, and the employment they obtained in Japan was often considered difficult and dangerous.

[§§] São João is celebrated throughout the north and northeast of Brazil on St. John's feast day, June 24, and is marked by the reenactment of a "shotgun wedding" in a typical rural village, with a predictable story line and much dancing. It is, in essence, a harvest festival.

[***] Cooperative membership and farm titles are typically held by men as heads of households. Several members emphasized in interviews that, despite a male-dominated membership, the cooperative served households, including female spouses, especially through the women's department, which had, among other things, experimented with supporting small-scale fruit preserves production.

CONCLUSIONS: LESSONS FROM THE CAMTA EXPERIENCE

This case study of CAMTA and its impact on economic growth, membership well-being, and environmental sustainability is, of course, unique. Nevertheless, it rather poignantly demonstrates the potential of smallholder cooperativism to fuel the development process in a sustainable way. The waves of Japanese immigrants who settled the banks of the Rio Acará came equipped with a tradition of cooperative action, a sense of ethnic solidarity and cultural "spirit" (*yamato damashii*), and a shared struggle. We argue that these qualities, channeled into a formal institution through the cooperative (after 1949), provided the mechanism of resilience and successful adaptation in the rain forest ecosystem.

The lessons learned from this case study are multiple. First, CAMTA was never only an economic asset to the membership. The cooperative systematically organized the search for new technological options—new crops, cropping systems, input provisioning, and so forth—and it defined its responsibility in terms of the broader needs of a community of members. The cooperative provided health care, education, and cultural and leisure facilities, as well as a youth group and a women's association—all of which contributed to the collective well-being of this community. Most of all, the cooperative became the central gathering point for community problem solving. As Piekielek (2010) has pointed out, the cooperative manages a pool of social capital as well as economic assets, and in this way, the cooperative has come to penetrate so many different areas of the daily lives of its members.

Second, the resilience of CAMTA in the face of so many financial and economic crises reflected a highly effective relationship with external actors, particularly the Japanese government (through JICA), Japanese and American private-sector firms, and the broader Brazilian policy-making apparatus. The cooperative showed itself open to new technology and new markets and was able to acquire capital investment support from Japanese and Brazilian development organizations and banks at critical times. Even today, as the agroforestry model of land use generates increasing attention at national and international levels, the cooperative actively opens its experience up to all interested parties and participates in information sharing at all levels. Such an "outward-looking" strategy has resulted in a constant stream of innovation and much institutional flexibility.

Finally, the CAMTA case study suggests that adherence to the fundamental democratic principles of cooperativism—bolstered by a shared vision and shared values, a strong sense of community, and a common struggle—can contribute to great advances in well-being. Contrary to the broader Brazilian context of sharp socioeconomic stratification, this cooperative in Tomé-Açu reduces differences and manifests the equity benefits of community problem solving and collective action. Not all members have the same level of resources and wealth, but all have an important role to play in determining the future of their cooperative and their community. There is a shared feeling of pride among the members in their common quest for sustainability—both in the sense of economic stability, the legacy that is passed on to their atotori, and in the sense of environmental sustainability, their successful integration into a fragile rain forest.

Sustainability is a dynamic process, not a static result that, once achieved, remains so. The cooperative will face challenges as it grows, most critically as the membership becomes more heterogeneous and the processing plant becomes increasingly dependent on raw materials produced by newer smallholder associations. Greater organizational complexity is inevitable as the membership extends to those who do not appreciate golf and baseball or the rich Japanese immigrant traditions that the community regularly commemorates. How the cooperative grows will require adjustments, particularly in strategies of problem solving. If the cooperative principles remain strong, such adjustments will be possible.

SHARECROPPERS TRANSFORMED

The Case of the Cooperativa dos Produtores do Curupati-Peixe (CPCP)

LUIS BARROS AND TIMOTHY J. FINAN

INTRODUCTION: DISRUPTION

IN 2002, the massive Castanhão Dam in the state of Ceará was completed, thus trapping the surface runoff of the Jaguaribe River Valley and its feeder watersheds into a vast reservoir (Açude Castanhão) capable of holding 6.7 billion cubic meters of water. The area that was flooded with the filling of the reservoir covered 325 square kilometers and completely inundated the city of Jaguaribara, the seat of the *município* with the same name.[1] As impressive an engineering feat as Castanhão is considered, the process of relocation and adjustment of the population was as daunting.

Ceará is one of nine states that form the northeast region of the country. The early Portuguese colony was established in the northeast, built around a sugar plantation economy along the relatively verdant coastal region. Toward the interior of the region, the environment is harsh—semiarid, mostly poor of soils, and subject to regular droughts. This drought-prone backlands region is called the *sertão* (its inhabitants *sertanejos*), and it evokes great meaning in the local imaginary as a place of suffering where resilience and the indomitable will of the population are the chief assets. The traditional economy of the sertão was integrated into that of the slave-based sugar plantation in that it provided food supplies, especially dried meat, draft animals, and leather to the sugar mills.

The relations of production that emerged in the sertão have been presented as "semifeudal" (Goodman and Redclift 1982) based on the social formations that emerged from a highly concentrated landownership. Large landholdings, called *fazendas*, were often owned by local ranching elites who maintained on their properties a skein of resident sharecropper families. Typically, a sharecropper received a residence and a plot of land (Johnson 1971) on which subsistence crops and, usually, cotton, were produced. In return for the land and residence, the sharecropper paid a proportion of production (usually one-third of food crops and one-half of cotton) to the landowner. In this system, the social relations between landowner and sharecropper were highly unequal and clientelistic, so that the sharecroppers turned to the landowner in times of crisis (e.g., drought), and the landowner could demand multiple forms of loyalty and support (Finan and Nelson 2001; Nelson and Finan 2009). In effect, the lives and livelihoods of the sharecropper and his household heavily depended on the largesse and good will of the *patrão*.

This background sets the context for the Castanhão Dam. Ceará has no naturally occurring surface flow, and access to water for humans and animals has always been a primary source of vulnerability. Over the last 40 years, the state has shifted from a rural society to an urban one. In 1970, approximately 65 percent of the population lived in rural areas and survived, albeit precariously, on rain-fed agriculture. Today 30 percent of the population is rural, and the state economy is based on tourism, industry, and services. One of the major concerns of statewide public policy is to guarantee urban supplies of water, especially to the city of Fortaleza, the capital and now the fourth-largest city in the country.

Begun in the early 1990s, Castanhão was key to the larger, well-planned transformation of the state—a modernization project that was premised on a reliable supply of water for an urban-based economy and the "rational" agricultural use of water to support the production of high-value crops in areas with appropriate soils. Carried out in the flush of the nascent Brazilian democracy, the Castanhão project became an example of participatory planning and public consultation, and it was widely and openly debated.

First of all, the reservoir would entirely inundate the urban center of Jaguaribara, the seat of the município of the same name. Over the course of nearly 10 years, the commission responsible for the Castanhão project brought in social scientists, architects, urban planners, and a variety of sector specialists to design the move of Jaguaribara town to higher ground and to manage the displacement of over 10,000 people. A replica of the local church was built in the new town *praça*, graves were exhumed and moved, new housing was built, and eventually the relocation of this urban population to Nova Jaguaribara was successfully completed.

The reservoir waters also extended out to flood fazenda homes, pastures, and sharecropper plots that dotted the rural landscape, disrupting the centuries-old livelihoods that had reigned in this part of the sertão. For this sizable rural population, several livelihood options were offered. The rural displaced who had known only the life of the sharecropper or smallholder farmer, cowboy, and agricultural day laborer were given a choice. They could remain as independent rain-fed farmers on their own plots of land (elsewhere on higher ground), produce cash crops on a plot within a new irrigation perimeter, raise dairy cattle, or cultivate fish on the edge of the new reservoir. This case study examines those sharecroppers who chose to raise fish.

A NEW LIVELIHOOD: TILAPIA

It was recognized in the planning of Castanhão that the families displaced by Castanhão backwaters would not have the expertise or the experience to engage in new market-based livelihoods requiring significant technological sophistication. Thus a network of technical advisors was contracted to provide the assistance to get the new livelihoods operational. In the case of the households that chose aquaculture, a well-trained expert from the State of Ceará Secretariat of Agricultural Development (Secretaria do Desenvolvimento Agrário—SDA) assumed the challenge of helping transform sharecroppers into market-savvy tilapia producers. The initial group, known as the pioneers (*pioneiros*), consisted of 10 former sharecroppers, and in organizational meetings it was determined that the group would register as a cooperative—Cooperativa dos Produtores do Curupati-Peixe (CPCP).[2] In 2003, CPCP was officially registered, and its members began to produce and sell fish—namely, tilapia—for Fortaleza and local urban markets. As of 2009, CPCP had become the only tilapia-producing cooperative in Ceará and had quickly emerged as a major regional supplier, producing 590 tons of tilapia per year to the benefit of 70 member families (SEBRAE 2009).[3]

While the cooperative and its membership have flourished, two factors distinguish CPCP from other cooperatives in Brazil. First, instead of a group of like-minded individuals pursuing the advantages and protections of a cooperative organization, the state, in the person of one highly committed technical expert from a government agency (SDA), convinced the pioneer members that a cooperative was the most beneficial organizational solution. The members had no previous training or experience with the Rochdale Principles, with the fundamental philosophy of cooperative action, or with the details of cooperative management and decision making.[4] Not only was the state the decisive stimulus and the source of technical knowledge, it also provided the initial capital to purchase equipment and set up a working capital fund. The Castanhão project also built a small community of stuccoed homes with plumbing and electricity, lighted streets, schools, and meeting centers.

Second, the technical expert introduced a collective production system, which in a clientelistic society is a notion as distant and vague as it is challenging. While labor-sharing practices (*mutirão*) among households have been historically common, there are few examples of collective production per se, where the benefits (and losses) are shared by collectivity. Equally novel was the concept of individuals buying "into" a collective organization and owning a share of it, so that individual rewards would be realized only if the collectivity were successful. In traditional clientelistic society, a sharecropper had no reference points or past models to make sense of collective organization. Survival in the sertão was dependent on a dyadic, individualist relationship with a person of superior power—the patrão.

Tilapia aquaculture requires a technology that is complex and demanding. The fish are raised in cages (*gaiolas*) organized in lines out into protected inlets of the reservoir, where the shallow fingers of water are constantly refreshed from the larger water body.[5] Production in 2003 was initiated with 15 gaiolas, a scale which has an annual yield potential of 30 tons of fish. By 2009 there were 588 gaiolas in the water. The fingerlings are purchased from a supplier located in a neighboring state, and they are placed in a certain type of cage and fed a specific ration carefully prepared in the cooperative warehouse. As the fish grow, they are placed into other cages and fed ration types and amounts consistent with this stage of the life cycle. After several months, the fish are retrieved from the cages and classified by size so that the finishing cages have fish of a uniform size and weight.[6] Those fish that are under the desired weight are fed a different finishing ration. At the time of harvest (about four months), the fish are recovered from the cages, cleaned, boxed, iced, and shipped to market. The buyers are present at the harvest with their trucks ready to receive the fish so as to maintain freshness. Small tilapia (under one kilogram) at the time of the research were processed into fillets by the member wives and sold to local schools for the school lunch program.

The collective production system defines a set of interrelated tasks—the ration managers (the "rations" team, in their parlance), who remain in the warehouse, and the boatmen (the "support/vigilance" team), who travel back and forth in small canoes, feeding the cages and handling the fish. This team is also responsible for the night watchman task. All members participate in the flurried harvest, which is an intense concatenation of activities carried out under the pressure of time. Not all tasks are equally demanding, so each member is rotated into a different role each month. The coordination of activities is critical for the entire system to function successfully. Ration types and sizes, distribution of feeding, monitoring of fish size, and maintenance of the gaiolas are all linked together into sequences and schedules, such that, if one person fails to perform a task at the scheduled time, the entire operation is put at risk.

The management of the system is equally complex. In 2009, only two of the members had a high school education, and they were responsible for the overall administration of the cooperative and for marketing. Not only are fish sold, but key inputs such as rations and fingerlings must be purchased on a regular basis. It is important to note that the transformed sharecropper never held such managerial decision-making responsibility in his previous livelihood.

Following cooperative convention, an elected board of directors is responsible for the decision making of CPCP and is composed of five members, including the president, the financial director, the marketing director, and two production directors. The president carries out the function of allocating, according to the member's capabilities, the different production-related functions. For example, if someone shows a particular talent for organizing charts and data, he would be assigned to the cage-organizing position; if he knows math, he could be assigned the task of keeping the accounting records of each sale. The marketing director must have some basic math/accounting skills to monitor and record the amount of fish sold. The finance director pays salaries and monitors the cooperative's net cash flow. And the two production directors manage the production activities. Regardless of administrative duties, all members belong to one of the two teams described above (rations and support/vigilance).

Board elections occur every two years with term limits set at four years (consecutive). Cooperative statute supports open membership, but in reality, the production system (e.g., the number of cages) imposes an unspoken limit on the number of members. Individuals are allowed to join based on a skill set that is needed (e.g., accounting knowledge) or based on social and kinship ties. A current member proposes a relative as a candidate and this candidacy is considered during a co-op assembly meeting. In the assembly, voting is not secret, and close ties of kinship are usually considered an important criterion. After all, the community itself is small, and everyone is essentially a neighbor.

During 2004, 15 additional members joined the cooperative, and by 2009, membership had grown to 70.[7] Under the collective production system, the costs and revenues are calculated monthly, a contribution is made to the capital reserve of the cooperative, and the "profits" are distributed equally as long as each member has fulfilled his task load. At the time of the research in 2009, the monthly return per member was approximately 1,100 Brazilian reais (US$500), well above the mandated national minimum wage. Members have seen the benefits of collective activity organized around the cooperative model and have expressed satisfaction with these results.

TRANSFORMATION TO COLLECTIVE ACTION: LESSONS LEARNED

A key question addressed in this volume focuses on the intrinsic tension of a cooperative model where economic and larger social goals seek a viable balance. In the case of Curupati-Peixe, the cooperative organization was burdened with an even greater challenge. A membership of mostly uneducated subsistence farmers whose traditional strategy of survival had been an individualistic and dependent relationship with their landowners was transformed into the owners/operators of a business enterprise that could only succeed through collective action. In effect, CPCP has thrived as an economic entity—it is profitable and has an effective business model; but the root of its success has been the acceptance of a cooperative form of production, where individual benefits are defined only by the positive outcomes for the enterprise. Through cooperative practices motivated by individual economic incentives, the membership has increasingly come to understand and embrace the Rochdale Principles that constitute the cooperative foundation. The realities of ownership, management, interdependence, and participation contributed to the transformation of the sharecropper into someone with both decision-making power and responsibility for the well-being of the totality. The cooperative imperative provided the instrument by which this transformation was effected.

A second lesson gleaned from CPCP was that success required strong, long-term support from the state. In the larger Castanhão modernization project, the original design envisioned groups of displaced sharecropper households organized under some sort of associative model and engaged in market-based productive activity. In this sense, the cooperative was imposed on this group of neophyte fish farmers. Nonetheless, even with the significant influx of start-up capital from the larger project, it is doubtful that CPCP would have gained purchase without the constant, almost persistent assistance provided by the SDA technical expert. The transformation of sharecropper to owner/operator would not have occurred without the intense incubation provided by the state. In most cooperatives, such as in the south of Brazil, the membership accumulates and manages its own assets and depends on the state for the enabling institutional framework and for episodic support in the form of credit lines, market opportunities, and so forth. In the case of CPCP, the state provided the capital, the collective production model, and the day-to-day technical orientation to move the group toward a sustainable business enterprise. Whereas in many places state intervention in the workings of cooperatives has been a formula for failure, here it appears to have achieved success.

CPCP faces numerous challenges as it adjusts to changing business climates and to its own growth, and these challenges will provide further lessons into the future. The first is for the cooperative to expand its share of an increasingly competitive market. The current marketing strategy follows a highly traditional structure where buyers appear at the gate to purchase fish. In 2009, two high-volume middlemen purchased four tons weekly for resale in the wholesale fish market in Fortaleza,[8] and two mid-volume buyers purchased around one ton per week of lower-grade tilapia for resale in weekly markets around the interior of the state. This market is vulnerable to oscillation in price, and the cooperative does not have reliable access to market information. It has considered the purchase of a refrigerated truck to transport its fish to market, thus taking greater control of the marketing process and eliminating some of the middlemen.

The membership has also discussed a diversification strategy to expand its product lines, as is the case of processed fillets for the school system. There is a large amount of waste by-product in fish innards, skin and bones (from the fillets), and so forth, and proposals for the production of processed fish cakes and the use of waste to raise hogs have been considered. The expansion of the product line is a form of spreading the market risk that concerns the membership currently.

The second and larger challenge is the need to develop management skills. While the co-op has taken advantage of technical assistance from the state, it must at some point develop its own management and administrative skill set in a sustainable and professional way. Currently, the membership does not have individuals with the professional administrative skills needed in such areas as accounting, information management, marketing, and finance. To achieve sustainability as an enterprise, the co-op must develop this talent internally or recruit it from outside. This lack of professional expertise will pose increasingly binding constraints on co-op growth.

The third challenge is to develop the social vision for the cooperative. This vision would include the projected size of membership and the strategy for new member recruitment. It is typical of the region that a father divides his assets among his children. In the case of the cooperative, a member cannot freely "bring" his children into the business because of the inherent limits of the co-op production. The intergenerational transfer of the tilapia enterprise to the next generation is a natural desire; however, the membership size is constrained by the scale of production, and it is not yet clear how the system can expand to absorb more members.

CONCLUSION: THE POTENTIAL OF COOPERATIVE ACTION

Curupati-Peixe is a success story for a local community of displaced families that documents the development of potential cooperative action. Its powerful message is that a model of cooperative production enabled a fundamental transformation of dependent sharecropper farmers into small-scale, empowered businessmen with new social identities and aspirations for the future. Regardless of past experience or status, each member recognizes the intrinsic value of his contribution to the collectivity and, in turn, understands how collective action enhanced the value of his labor. The membership has come to understand that working together in a coordinated way can generate a collective good in which all members participate. This will be a lasting legacy of Curupati-Peixe—cooperative and community. CPCP will face obstacles in the future as it negotiates an uncertain capitalist system that demands complex knowledge of value chains, input and output markets, production technology, and management skills. Fortunately, the membership is aware of these challenges and has grown in its decision-making sophistication. And for now, the co-op and its membership can take pride in the great strides already recorded.

NOTES

1. In Brazil, a município is the small politico-administrative unit of government. A município typically has an urban center surrounded by a network of small settlement clusters in the rural areas. In terms of size, municípios can vary widely, from less than 100 square kilometers to more than 4,000 square kilometers, and the state of Ceará has 184 municípios.
2. Displaced households who sought other livelihoods, such as irrigated production, also formed cooperatives: Curupati-Irrigação (irrigation), Curupati-Leite (dairy), and so forth.
3. The analysis that follows is based on data from Barros's two-month stay in the cooperative community, interviews with technical experts and fishmongers in the Fortaleza, and follow-up communication with the community.
4. Nor did they know one another for the most part. The families tended not to be related or to share much of a history.
5. The original site for the gaiolas proved inadequate and, in an environment of crisis, a new, more appropriate site was selected.
6. This selection process is called *repicagem*, a term that in horticulture refers to transferring selected plants from a seedbed to another site where the seedling will grow to transplant size.
7. It is also true that the market for tilapia expanded rapidly during the period throughout the state, providing the cooperative with the incentive to increase membership.
8. The Carlito Pamplona market in Fortaleza is itself iconic. It is a sprawling set of booths that operate during the early hours of the morning to sell to fish retailers. There are several varieties of fish sold in this market, mostly freshwater fish, and sellers may come from all over the northeast. It is a highly volatile market.

SOLIDARITY, TENSION, AND CHANGE ALONG THE AMAZON RIVER

The Case of ASCOPE

TIMOTHY J. FINAN

THE RECURRENT QUESTION in this book is, how exactly does cooperative organization confer economic advantage on a group of rural producers and offer them a pathway of dynamic change in community well-being? The conventional explanations create a narrative of market strength and economies of scale at various stages of the commodity chain; but in highly stratified rural societies with unequal access to capital, technology, and information, such advantages seldom occur. The foundational Rochdale Principles of cooperativism presume both essential and operational equality of participation and voice, and where such equality is not present, the cooperative as institution adopts the structures of inequality of which it is part. This case study, however, presents a success story of cooperative action. The community described here—Comunidade do Sagrado Coração de Jesus do Paraná da Eva (SCJ)*—is located in the state of Amazonas in the middle of the Brazilian rain forest, and its cooperative, ASCOPE (Associação Sagrado Coração de Jesus do Paraná da Eva), provides an example of cooperative-led growth and development seldom documented in the cooperative literature. This story is about how a group of producers coalesced around a common ideology of justice and

*"The Community of the Sacred Heart of Jesus of the Paraná da Eva"—a *paraná* is a water body resembling a cut in a feeder river.

self-sufficiency and succeeded in creating a burgeoning local community built upon a sustained market economy.

One clear overriding fact about this community is that it has changed in highly significant and measurable ways, including population growth, rapid increase in diversity and intensity of economic activity, large public investment in infrastructure (asphalt roads and electricity), and the introduction of essential public services (schools and clinics). These changes occurred over fewer than 20 years, and they are strongly associated with the success of ASCOPE as both a community project and an agribusiness.

The history of the community is deeply intertwined with that of the cooperative, and to those familiar with Amazonian rural society and its many forms of exploitation of both natural resources and people, the outcome of this story appears an improbable one. The chapter, based on field interviews from 2005 and 2009, documents the founding and growth of both SCJ and ASCOPE and explores the moments of crisis when the continued viability of the cooperative required significant reflection and difficult decisions. The great and constant challenge faced by both community and cooperative membership has been to mediate between the ideological commitment to solidarity and the realities of market and capitalist entrepreneurship. How this group has negotiated the inherent tensions between these two paradigms of development holds lessons for rural producers everywhere.

THE SETTING

The immense Amazonian floodplains, called the *várzea*, have supported human livelihoods for 12,000 years. Since European settlement in the region, and since the eighteenth century, European, indigenous, and mixed native-European (*caboclo*) communities have occupied these river edges of the Amazonian basin engaging in subsistence agriculture, livestock, fishing, and extraction activities.† Annually the rivers spill their banks, covering nearly 100,000 square kilometers of low-lying forests and depositing nutrient-rich sediment brought from upstream. The water-level difference between the terrestrial and aquatic phases of the várzea can be as much as 10 meters. As the floods recede, local

†In the middle half of the twentieth century, the várzea was also a site of widespread jute production, an agro-industry now in decline.

residents plant subsistence crops and manage the emergent pasture grasses, fish and hunt, and collect nontimber forest products for consumption and sale. The várzea is a complex ecology to which equally complex livelihood systems have adapted over the generations.

Most households in these várzea communities are land scarce, with limited access to public services such as education and health care. Although recent federal programs have brought electricity to most towns, transportation and communication infrastructure remains underdeveloped. Houses are built above ground on stilts at the edges of the rivers, lakes, and *igarapés* (freshwater lagoons), and the main source of daily transportation is canoes powered by small electric or gasoline motors (*rabetas*). It is a mostly remote and waterbound life, defined by the seasonal rhythms of this vast riverine system.

Sagrado Coração de Jesus do Paraná da Eva is located 227 kilometers east of Manaus, the capital of Amazonas State, and is part of the *município* of Itacoatiara.[‡] The community itself now has about 250 families (Souza and Jesus 2011), and its buildings were constructed above the reach of the seasonal floodwaters, on *terra firme*, or the "high ground" away from the flood plain.[§] Recent history indicates that the first named community in the area of the Paraná da Eva was called Vila do Engenho, located around the Lago do Engenho (Jesus 2000). This settlement, established in the 1950s with the arrival of the founding family, was dispersed around a sugar mill (*engenho* in Portuguese, thus the name), the abandoned ruins of which are still to be found on the edge of the Lago. The population of the Vila do Engenho had mostly migrated from other parts of Brazil, particularly from the drought-plagued and land-scarce northeast of Brazil and from neighboring parts of the várzea (Jesus 2000). Livelihoods along the Lago were highly diversified, combining small-scale production of food crops and jute in the várzea (during the recession of the waters) with the raising of cattle and chickens, then moving to temporary upland housing during the floods. These farm families were often landless or occupied várzea lands illegally (*posseiros*), and there was frequent tension between these farmers and

---

[‡]The município in Brazil is the local politico-administrative level of government. The *prefeito(a)* is the locally elected executive who runs the *prefeitura*, roughly, the "mayor's office."

[§]Of architectural interest perhaps, many houses in the community are still built set off the ground even though riverine flooding is no longer a risk.

the large-scale ranchers (*fazendeiros*) who raised their animals on the pastures of the várzea.

In the 1970s, the Catholic diocese in the município seat of Itacoatiara organized the community that is today called Sagrado Coração do Jesus do Paraná da Eva. The residences were constructed on terra firme overlooking the river near the Vila do Engenho, and the new community was inhabited by two dozen families who shared a commitment to the religious and social values of liberation theology—primarily the importance of justice and solidarity. From an analytical perspective, it is important to emphasize that SCJ represented the coincidence of physical and social space. To an outside observer, SCJ was indeed a small nucleus of houses organized along linearly planned streets, but to the residents it was also a well-defined and active social construction with solidarity and equality as the brick and mortar.

This basic fabric of shared religious and social values was further supported by the emerging political influence of the Workers' Party (Partido dos Trabalhadores—PT) in the município of Itacoatiara. The theme promoted by the political leadership and the local Farm Workers' Union (Sindicato de Trabalhadores Rurais) was that the SCJ rural *workers* (*trabalhadores rurais*) would undergo a transformation into rural *producers* (*produtores rurais*) fueled by an extremely powerful narrative of strength in solidarity and collective action.

In 1980, a charismatic political leader who had helped found SCJ led the effort to create a small group of farmers, called Grupo Lavradores em Ação (Farmers in Action), or LA, in order to put the essential principle of solidarity into action. There were 16 families from SCJ in the initial membership of LA; nine of them were families of landless day workers. Thus, the first challenge was to obtain land, which the group successfully did through various mechanisms, including the granting of parcels by landed members. Steadfastly against the perceived evil of salaried labor, the group employed two strategies of labor sharing in the production process. First, to prepare the (forested) terra firme land for cultivation, a large contingent of members (called the *grupão*, or "large group") went from parcel to parcel working together to clear the land and ready it for cultivation. Once this task was completed, every three families were organized into a *grupinho* ("small group"), which shared the labor tasks of production (planting, weeding, spraying, harvesting). In this upland, the principal cash crops were pineapple (*Ananas comosus*) and the tropical fruit *cupuaçu* (*Theobroma grandiflorum*).

Not only did the LA see exploitation and injustice in the traditional relations of production, they also perceived the marketing system as a form of exploitation, both in the provisioning of basic consumer goods and in the purchase of farm output. Accordingly, they formed marketing teams consisting of community members, who took the pineapples and cupuaçu to markets in Manaus for a group-approved fee.

In 1990, after a decade of active production and cooperative learning, the Lavradores group then decided to register as a formal cooperative under the current name, ASCOPE. The members realized that as a cooperative, they would be eligible for public financing destined for the support of small-scale agriculture. With the first bank loans, the cooperative set aside half the amount as a guarantee of repayment and used the rest to purchase a floating general store (*flutuante*) anchored riverside on the Paraná da Eva. This store, still thriving and popular, provided basic food and nonfood items to consumers—members and nonmembers—at prices lower than other local businesses. In accordance with basic market theory, the flutuante introduced new competition into the area, driving down the prices of rival businesses; but it also became an active point of informal social gatherings with small boats and their rabetas pulling up to make weekly purchases and catch up on news and gossip.

What SCJ did not have in its newly established community of the 1990s was electricity and infrastructure. There was limited schooling available to children, and health services were offered only in Novo Remanso, a town located about 30 kilometers away but with poor access due to the precarious roads. If the sense of solidarity dulled the edge of need, the community was nonetheless a remote and poor one.

FROM SOLIDARITY TO MARKET CAPITALISM

The trajectory that led from várzea to terra firme, from worker to producer, and from Lavradores em Ação to ASCOPE mounted a business model on an ideological foundation of solidarity and justice, which has been presented as an example of the "economy of solidarity" (Jesus and Jesus 2006). In part, such solidarity came "naturally," in that it was well enmeshed in the cultural fabric of the small community that created the two groups. The 16 members that formed Lavradores em Ação came mostly from two large families, the Peixotos and Barbosas, who had intermarried and were mostly all related to each other through biological or affinal bonds and through the common cultural tie of godparenthood.** This sense of community, however, needed to demonstrate an attractive economic outcome.

In 2005, the SCJ community had around 120 families, and ASCOPE had 30 producer-members. In 2013, the official population of the community (including the original Lago do Engenho) was 837 families with 230 families living in the "urban center" or the "agro-villa" (the original area of terra firme where the first settlement was constructed). Most of the families in this region make their living off agricultural production. The membership in the cooperative had increased to over 50 producer-members by 2009, with the prospect of further expansion. The agricultural livelihood is based on the production of pineapple and, to a lesser extent, cupuaçu. In 2005, there were an estimated 500 hectares of cupuaçu planted and 800 hectares of pineapple (Jesus and Ribeiro 2008), although this ratio has probably shifted as the result of disease problems in the cupuaçu and the increased market access of pineapple. In 2005, ASCOPE sent an estimated 20 million fruits to the market in Manaus, and by 2009, this number had risen to 36 million fruits.††

The production technology employed by ASCOPE members is advanced. Field sizes are less than five hectares, with around 20,000 plants per hectare in double rows. The use of chemical fertilizers and pesticides is part of the cultivation practice, as is the timing of plant maturation using calcium carbide to hasten flowering. In this way, the supply of fruits can be planned over the life of the plant. When harvested, the pineapple fruits are stripped of their slips and suckers, which are saved for replanting. At the end of the production cycle, the plants are removed and the field is prepared for the next cycle. In the case of cupuaçu, the district of Novo Remanso (of which SCJ is a part) is the largest producer in the state of Amazonas, with around 1,000 hectares in cultivation and a production of somewhat more than 2 million fruits in 2011

**In Brazilian culture, the godparents of children are socially, although not necessarily equally, connected in a strong social bond that involves common responsibilities between the two sets of adults.

††The Amazonas Institute for Agricultural and Forestry Development (IDAM) has estimated that, in the district of Novo Remanso, around one thousand family farmers cultivated 3,000 hectares of pineapples and produced over 60 million fruits in 2015. This suggests that the pineapple "fever" has been more widely adopted by farmers in the region.

(IDAM 2015). Cupuaçu is related to the cocoa family and is a perennial. Originally, ASCOPE farmers intercropped the young cupuaçu in pineapple fields, but since the young plants require shading, they are now commonly grown among other tree species.

Along with the intent to eliminate daily wage labor as a rural livelihood, ASCOPE sought to create its own marketing chain in order to reduce the production value captured by intermediaries. It created a marketing group that charged a fixed amount to sell pineapples in the central market of Manaus. Later, however, the management of ASCOPE entered into a contractual agreement with the município and state secretariats of education (SEMED and SEDUC) to deliver pineapple and cupuaçu to the school feeding programs in Manaus. Brazil has a highly progressive school lunch program whereby the federal government subsidizes the purchase of local foodstuffs to provide lunches for all students in the primary and secondary school systems. In addition, the cooperative participates in several farmers' markets in Manaus that are sponsored by the large military base (CEGIS) and the "modern" open-air market in the city.[‡‡] This marketing strategy has significantly expanded the name recognition of ASCOPE products and has reduced instability in prices.

In 2002, ASCOPE received government financial support to purchase an agro-industrial plant with refrigeration capacity, trucks, and a tractor. In 2005, the cooperative obtained land to install the facility, and in 2009, after electricity and a road arrived, the pulp-processing plant began production. According to the state and federal organizations that provide technical support to cooperatives (OCB-AM, SESCOOP),[§§] the pulp plant marketed 100 tons of cupuaçu pulp and 80 of pineapple pulp in 2012, nearly all of it to Manaus, although efforts are now in progress to reach European fruit pulp markets.

ASCOPE is widely promoted throughout the state (and more broadly) as the paragon of cooperative development. Based on the comparison of two points in time (2005 and 2009), the growth of the SCJ community and the

[‡‡]CEGIS is the largest school for tropical guerrilla warfare in Brazil (and perhaps in Latin America), and it sponsors a large biweekly market for locally produced products on the base open to the public; the "*Mercado Moderno*," sponsored by the state government, is an open-air market consisting of producer booths with a wide variety of fresh foods and a large public following.

[§§]OCB-AM is the Organização Brasileira para Cooperativismo—Amazonas, part of the national OCB cooperative-support network; SESCOOP is Serviço Nacional de Aprendizagem Cooperativista, the federal cooperative-learning organization.

diversified expansion of its economy provide remarkable signs of vast improvements in community well-being. In 2009, there was a small hotel, a restaurant and several luncheonettes, a pharmacy, a supermarket, and multiple other commercial points. The town has a paved road to the main Manaus-Itacoatiara highway, electricity, and Internet access. While there was not a resident doctor in 2009, there is a health post for malaria testing and a community health worker. In 2004, SCJ also received a school that serves grades six through nine for the surrounding region, as well as high school and evening adult education. As important as the inventory of infrastructure and services, SCJ has a very strong feeling of community, and there is ample evidence of collective action—the evening activities at the sports complex next to the school, decorated streets and painted reminders of upcoming commemorations and events, and, notably, families and friends of all ages walking the evening streets and sidewalks.

SOURCES OF TENSION WITHIN ASCOPE

Growth with equity is an age-old development challenge, and in the case of ASCOPE, there is an inherent tension between the logic of market capitalism and the community values of solidarity and justice. The cooperative has accumulated sizeable assets, expanded its market reach, diversified into agricultural processing (i.e., fruit pulp production), and intends to increase its product line into other exotic fruits. It operates a retail consumer and supply business, markets its own products, employs a number of local workers, organizes the distribution of collective assets, such as the cooperative tractor, and maintains multiple networks with external actors from banks to government agencies to private-sector representatives. In 2005, cooperative accounts were kept at the manager's home by his wife; now full-time professional assistance is required. Under the demands for greater skill specialization, larger payrolls, and more complicated accounting, administration, and management, it is predictable that a distance might form between cooperative leadership and the membership. ASCOPE does not have the resources to contract professional management, so the decision- and policy-making role is blended with that of the manager.***

***Under most cooperative structures, the board of directors (elected by the membership) contracts a professional manager (an employee) to carry out its policies and decisions.

The concentration of decision-making power presents the danger of creating an administrative "elite" less responsive to the need and priorities of the membership and to the values of cooperative action that bind them.

This dynamic tension between the requirements of a growing capitalist firm and the solidarity emerged from structural differences within the membership. Among the 30 original members, only two had finished high school, while half either had no formal education or had not finished primary school. There were also among the 30 clear differences in income and wealth levels as measured in assets (land, cattle, cars, etc.). In 2005, some members stated that they remained silent during meetings and other public activities, feeling that their lack of education reduced the value of their comments and opinions. In a similar vein, others insinuated that they felt intimidated by those who dominated the flow of the meetings.

Thus, while the origins of ASCOPE lie in strong values of solidarity and social justice, the management demands of the ASCOPE-cum-enterprise have seemed to distance cooperative leaders from the membership, creating an ideological division. During interviews in 2009, it was apparent that management had been increasingly perceived as being more involved in commercial business affairs and external negotiations at the expense of solidarity and social learning. Some original members felt excluded and stepped away from the cooperative or reduced their participation. To its credit, ASCOPE management addressed this basic identity crisis and sought to reincorporate disaffected members and reinforce the founding principles; and the current charismatic director/manager proudly points to the fact that several adult members, including an ex-director, have decided to return to school. Nonetheless, this tension is an inherent characteristic of the challenge to reconcile basic economic and social goals, and it appears unavoidable.

Other impending challenges are a natural consequence of ASCOPE's success. On the economic side, the cooperative will increasingly encounter technical capacity constraints that can be binding at the level of production (e.g., changes in cropping patterns, response to pests), at the processing level (e.g., equipment maintenance, quality control), and at the administrative and marketing level (e.g., volatile markets, policy impacts). The cooperative has to either develop this technical capacity internally within the community or access it from external sources. In the past, the ASCOPE management has been very effective in its ability to attract public funds and technical assistance, but such ready availability of outside capital and knowledge carries no guarantee into the future.

On the social side, ASCOPE must address the size of the membership, how new members will be recruited and under what terms, and how their incorporation will continue to maintain the sense of solidarity and community so important to the sustainability of the cooperative. This vision of the future requires a reflection on limits—including those implied by community commitment, a shared set of values, and even the environmental limits of the land and forests.

SOLIDARITY AND DEVELOPMENT: LESSONS FROM ASCOPE

Students of Amazonian rural society (e.g., Jesus 2000) propose that the unique nature of the várzea–terra firme ecosystem shapes livelihoods in a way that favors deep community bonds. The "shared isolation" of a waterbound existence engenders a strong psychological dependence on neighbors and a need for a community identity. Certainly, the history of Comunidade Sagrado Coração de Jesus, with its founding family lineages, conforms to that narrative. What makes SCJ unique is that it engendered ASCOPE—an active participant in market capitalism but predicated on a community vision of solidarity and social justice. This chapter has explored how this merging of community and capitalism can have positive development outcomes, measurable in terms of well-being and quality of life.

Several forces united to make ASCOPE a successful experiment in collective action. There can be little doubt that ideology of class consciousness and social equality was the enabling condition for the establishment of, first, Lavradores em Ação, and subsequently, ASCOPE. The hours spent as a group at a member's home discussing and reflecting upon the meaning of justice, the principles of associativism, liberation theology, and political participation created a process of social learning that gave meaning to the cooperative vision. More than ideological harmony, however, the ability of the cooperative leaders, albeit without much formal education, to transform the ideology into praxis put the cooperative on a pathway to success.

The ASCOPE leadership, with the strong support of the local political structure in Itacoatiara (the Worker's Party) and the Church (a diocese steeped in liberation theology), very actively took advantage of public policy incentives from a range of state and federal agencies. Agricultural lands were acquired through INCRA (the federal agrarian reform agency); the cooperative received public financing from the FNO (Fundo Constitucional do Norte), a federal development fund; the fruit pulp processing plant with its refrigeration capacity was installed in SCJ with public-sector support; and ASCOPE has received regular technical assistance from IDAM, the state agricultural extension agency, and from OCB-AM, the state agency of the national system that provides cooperative assistance (Jesus 2000). Over the last decade, the cooperative successfully negotiated contracts with the prefeitura of Manaus and the state of Amazonas to provide fresh fruit and fruit pulp to the compulsory school lunch program and gained access to several urban venues for the direct weekly sale of the SCJ product line. As a consequence of this history of public-sector interaction, ASCOPE has become well-known and respected throughout the state and the entire Amazonia region as a model of agricultural success.

ASCOPE has also benefited from a charismatic and energetic leadership that promotes the founding principles and values of the cooperative but is also highly pragmatic and effective in its business practices. Few members of the cooperative travel to Manaus, the large urban capital, but the cooperative leaders are frequently present in the market places, the government offices, and the OCB and SESCOOP facilities and classrooms. In a sense, the leadership lives in two worlds and must regularly negotiate the underlying contradictions between these two realities.

Ultimately, the success of ASCOPE depends upon the willingness and ability of farmers in the Comunidade Sagrado Coração de Jesus to produce pineapples, cupuaçu, and other fruits that have market value. In 2009 interviews, a sample of these farmers articulated that their goals are family-centered—they desire to increase the return on their labor in order to improve the opportunities for their children, achieve a sustainable economic living into the future, and maintain the self-respect that comes with control over their own destiny. As readily, however, they see collective action as the pathway to these goals. The continued viability of ASCOPE, therefore, is as much linked to this shared conviction that solidarity generates equal opportunity—that social justice and collective action are instruments that feed individual ambition—as it is to the expansion of markets, acquisition of agro-processing equipment, or the reduction of marketing costs.

ASCOPE will continue to reconcile the inevitable tensions within its group. New members will be needed to address generational transitions and demographic growth as well as to provide the pulp plant with raw materials and markets with product. The cooperative will require ever-increasing levels of complex technical skills in agricultural production and processing, in administrative professionalism, and in marketing. How these challenges are met will define the sustainability of the cooperative, but the likelihood of success will mostly depend upon constant reaffirmation of the human dignity of each individual, the value of solidarity and justice, and the continued emphasis on social learning, group reflection, and collective problem solving. In short, the history of a seamless cooperative-community must also be its future.

BRINGING POWER AND HISTORY INTO COOPERATIVE DESIGN

Lessons from AmazonCoop's "Fair Trade" Partnership Among Indigenous Brazil Nut Harvesters and The Body Shop

BRIAN J. BURKE*

AS ORGANIZATIONS COMMITTED to radical social transformation, cooperatives face a serious challenge: they must build on the present in order to change it. To succeed, cooperatives must strike a fine balance. Those that cannot meet the demands of highly competitive markets risk financial collapse, but those that simply conform to the status quo become complicit in re-creating inequality and marginalization. Faced with this dilemma, many cooperative managers and commentators have prioritized pragmatic questions of how to launch and sustain cooperative economic projects. In the process, however, they often ignore questions of history, power, and change that should be central to cooperativism.

In this chapter, I examine AmazonCoop from a historical-political-ecological perspective in order to demonstrate how a thorough consideration of power and history can enhance cooperatives' impacts as deeply transformative change agents. A power-centered approach directs our attention to the noblest goal of cooperativism—which is, after all, not to establish a particular type of

---

*A previous version of this chapter was published in the *Journal of Political Ecology*, vol. 19, paper no. 9. The empirics and arguments in the two pieces are nearly identical. However, the version printed here has been reduced in length, in large part by eliminating a discussion of how AmazonCoop is part of a more general effort to promote Amazonian forest conservation through community-based development.

organization or survive in a global economy, but rather to create more just social and economic arrangements—and suggests innovative possibilities for achieving this goal.

AMAZONCOOP

AmazonCoop (also known as CAMPEALTA) was founded in 1998 by representatives of the Fundação Nacional do Índio (Brazilian National Indian Agency—FUNAI) and the UK-based cosmetics company The Body Shop to formalize trade links between the multinational firm and Brazil nut harvesters from eight indigenous tribes. The cooperative was intended to supply The Body Shop with the raw materials and ecologically sustainable, socially just symbolism needed for its best-selling products, to help FUNAI protect and serve indigenous people despite neoliberal reforms that decreased funding and restricted their responsibilities, to allow indigenous people to finance their own community development, and to protect the environment by offering tribes an income-generating alternative to alliances with mining and logging companies. As a public-private partnership based on fair trade, the cooperative grew directly out of processes of neoliberal economic globalization and movements to harness globalization to benefit poor and excluded people. The cooperative was gradually destroyed between 2006 and 2007 as a result of legal troubles, strained relations with The Body Shop, and serious power struggles among the cooperative's nonindigenous stakeholders. Nonetheless, it offers valuable lessons for other cooperatives.

AmazonCoop was located in Altamira, Pará, in the central region of the Brazilian Amazon.[†] Although Altamira was important in the rubber booms at the opening of the twentieth century and during World War II, the local economy is currently supported primarily by legal and illegal logging, agriculture, and the extraction of nontimber forest products such as Brazil nuts, which are the fifth most valuable nontimber forest product nationally (IBGE 2004). Intensive natural resource exploitation has resulted in severe deforestation, water pollution, and the invasion of indigenous territories (Fausto 2001, 99; Fisher

[†]Amazonia is a culturally, ecologically, and economically diverse region that defies easy generalization. When I refer to the Amazon in this paper, the modifiers "Brazilian" and, in some cases, "Altamiran" are implied.

2000). Nonetheless, satellite images showing relatively intact forest on indigenous lands (Schwartzman and Zimmerman 2005; Zimmerman et al. 2001) and ethnoecological studies (Posey 2002) have stimulated efforts to preserve Amazonian ecosystems through indigenous development projects similar to AmazonCoop's Brazil nut trade. Outside of these programs, most of the region's indigenous people combine subsistence activities with cash incomes from mining and logging concessions, employment in extractive industries, and government pensions (Castro 1992).

AmazonCoop generated income through Brazil nut sales, an Internet service provider, and an eco-lodge in order to provide social and economic benefits to more than 2,000 indigenous people living in 14 villages and representing 8 indigenous tribes: the Arara, Araweté, Asurini, Curuaia, Kayapó, Parakanã, Xikrin, and Xypaia.[‡] Membership in the cooperative was organized at the village level, with one or two representatives from each village granted the members' right to vote in the General Assembly, but these village representatives were not elected and the vast majority of decisions were made by international advisors and nonindigenous Brazilian managers. As I have discussed elsewhere (Burke 2006, 2010), the cooperative's social and economic impacts in indigenous villages were inconsistent:

> AmazonCoop's international alliances certainly provided significant material benefits. The Brazil nut trading program more than doubled the annual profits of indigenous nut collectors while steering tribes away from alliances with the more environmentally damaging mining and logging industries, and tourist-funded, cooperative-managed health and education projects further contributed to a widely supported form of development. The cooperative provided indirect benefits by supporting [government services, both financially and administratively]. Unfortunately, AmazonCoop also made indigenous people more vulnerable and more dependent on outsiders, failed to include them as informed participants in their own development, masked the negative effects on them of unfavorable government policies, and perpetuated discriminatory distinctions among them. (Burke 2010, 42)

[‡]For consistency, I use the same ethnic categories as AmazonCoop. The Xikrin, however, are considered to be members of the larger Kayapó ethnic group (though they are somewhat distinct from other Kayapó groups), and the degree of involvement of the Xypaia and Curuaia is unclear.

Our initial analysis of AmazonCoop, based on research conducted in 2005 (Burke and Canova 2006; Burke 2006), demonstrated that global markets and international alliances provided significant economic opportunities for the cooperative and its indigenous members through increased incomes and support for new infrastructure. However, it also showed that successful cooperative economies guarantee neither genuine cooperativism, nor social development, nor transformation of the structural conditions that maintain marginalization. In particular, our research raised serious concerns about AmazonCoop's structure and its success in altering the deep inequalities that plague Brazilian society. Cooperative managers and members failed to translate the principles of democracy, participation, autonomy, and self-sufficiency into action, leaving indigenous people in the region as passive beneficiaries subject to the decisions and desires of the cooperative's international advisors and the local, nonindigenous cooperative staff, including the director of FUNAI-Altamira. This lack of indigenous participation raised concerns—which were, unfortunately, borne out—about the cooperative's sustainability.

In the present chapter, I consider the cooperative from a slightly different perspective. By examining AmazonCoop in the light of indigenous people's involvement in historical extractivist economies in the Brazilian Amazon, I analyze how the cooperative altered the regional political economy and the sources of power and patterns of exploitation of indigenous people. To conclude, I argue that cooperatives can act most effectively as agents of change only when their activities are explicitly designed to alter historically unequal power relations.

HISTORICAL STRUGGLES IN AMAZONIAN EXTRACTIVISM

The contemporary globalization underlying AmazonCoop's work has deep historical precedents in the Brazilian Amazon. For almost four centuries, the resource-rich Amazon has been a source of wealth for colonial and capitalist systems, and the resulting extractive industries have been important sites of political and economic negotiation among Amazonian indigenous people, nonindigenous Brazilians, local and foreign commercial interests, and the state. Understanding the impacts of AmazonCoop and contemporary globalization requires that we investigate the history of Amazonian extractivism, and particularly the history of the rubber industry. Three features of the rubber industry are particularly important for this discussion: the involvement of indigenous people in the rubber industry, the organization of rubber extraction through the *aviamento* debt-peonage system, and the recurrence of the aviamento system in Amazonian political-economic organization throughout the twentieth century (Burke 2006).

The Amazon presented would-be rubber barons with a number of serious obstacles. As Bunker (1984) explains, each mode of extraction in the region has been shaped by the social and ecological legacies of previous extractive projects. The late nineteenth-century elite could not easily respond to nearly insatiable rubber demand because the colonial-era spice and animal oil trades had degraded ecosystems, decimated the local, indigenous labor pool, and encouraged survivors to flee to upland areas from which they could more effectively control contact with Euro-Brazilians. To capitalize on new global markets and the high quality of Brazilian rubber, the elite were forced to import tens of thousands of workers to the region, mostly from drought-burdened Ceará.

If importing laborers was difficult, ecological and economic barriers posed even greater problems. *Seringalistas* (rubber bosses) struggled to make a profit in a highly volatile world market while using imported, impoverished, and inexperienced laborers to collect widely dispersed wild rubber in "what may be the most inaccessible area in all of Brazil" (Nimuendajú 1920, 149). They faced high costs related to river transport, numerous intermediaries between tapping and export, and the provisioning of tappers so that they could dedicate themselves to rubber collection rather than subsistence. They also faced high risks of losing their product on dangerous rivers, of losing upward of 50 percent of their tappers to disease and desertion, of losing profits due to unfavorable exchange rates and rapidly changing prices at export houses, and of being cheated when workers covertly sold rubber to itinerant traders (Barham and Coomes 1994).

The infamous aviamento system effectively overcame these obstacles to economic organization. Through this system, rubber tappers were transported into the Amazon and advanced supplies at the beginning of each season with the expectation that they repay these debts in rubber. Seringalistas set the terms of trade and in many cases used extreme violence to ensure payment. Minor seringalistas entered into similar arrangements with larger ones, and larger ones with the export houses that ultimately controlled prices and credit. Tappers, located at the bottom of the debt and pricing pyramid, paid as much as 250 to

500 percent above city prices for basic goods and received less than 50 percent of going prices for their rubber (Barham and Coomes 1994).

Indigenous Amazonians were profoundly, though not uniformly, affected by the rubber boom, particularly as tappers invaded upland areas in search of *caucho* trees and forced indigenous populations to flee, resist, or join the rubber economy as willing and unwilling laborers. Some tribes in the Altamira region fell prey to disease and a "slow and cruel war of extermination" (Nimuendajú 1920, 146). The Kayapó-Gorotire, for example, were attacked by rubber tappers after a period of peace and forced to migrate several times, initiating a long-standing war against the tappers of the Xingu, Fresco, Iriri, and Curua Rivers. Widespread violence made the Gorotire so haunting in popular mythology that Fisher (2000) reports present-day Brazilians shouting "Gorotire" to scare their children into good behavior. Other tribes worked with rubber tappers to identify trees, defend territories against competitors, and provide food in times of need, all in exchange for manufactured goods (Fisher 2000). The unequal influx of weapons affected intertribal power dynamics, forcing some tribes to migrate away from their well-armed neighbors. Enterprising seringalistas often exploited intertribal tensions, arming and paying "pacified" tribes to attack and sometimes enslave those who posed a greater threat to extractivism (Nimuendajú 1920, 152). A different Kayapó group, however, "had the rare fortune of making their first contact with well-intentioned civilized people . . . in 1891," and thereby "escaped from extermination by iron and fire, which was the fate of the others during the great advance of the *caucho* rubber tappers in this region" (Nimuendajú 1920, 147). Their good luck was not the norm.

Nimuendajú described, north of the Kayapó, other "leftovers of tribes, meager wastes that the wave of caucho rubber tappers did not drown in its brutal advance" (1920, 150). The Asurini "appeared" in 1894 and immediately launched a war against the "civilized people," maintaining an area between the Xingu and Pacajá Rivers by attacking on both sides. But the seringalistas armed and paid the "pacified" Arara to attack the Asurini, reducing their numbers and power significantly. Nimuendajú described the Arara and the remaining Xipaya and Curuaia as "belonging to" different white settlers, a relationship that he says saved the latter groups from the full onslaught of the rubber tappers but also forced some Xipaya into "conditions that can only be clearly characterized by one Portuguese word: slavery" (Nimuendajú 1920, 152). The Juruna, at one time the most important tribe of the Xingu region, were attacked, forced to migrate, and attacked again before finally establishing peaceful relations with a rubber tapper who took several leaders to Altamira. Most of the travelers died in the city, and the survivors fled into the forest.

After the Brazilian rubber industry collapsed under competition from more efficient Southeast Asian plantations, many indigenous people fled to more secure upland areas, and nonindigenous people migrated in search of other economic opportunities. But the aviamento system continued to organize Amazonian labor and commodity flows, including the extraction of Brazil nuts, throughout the twentieth century (Laraia and da Matta 1967). Aviamento remained a "rational" way of organizing and controlling scarce labor and capital in an Amazonian ecological context, and had likely become a fairly hegemonic template for social and economic organization. Relations among regional elite and between the elite, intermediaries, and extractivists certainly changed over time (see, for example, the discussion in Weinstein 1983), but one hundred years of patron-client networks organized around the aviamento system likely made it difficult to conceive of viable alternatives, let alone establish them against entrenched elite interests.

The case of the Kayapó-Xikrin demonstrates the continuity and institutionalization of the aviamento system and the importance of extractivism in regional political economies. After years of disrupting local extractive efforts, the Xikrin were "pacified" at the end of the 1950s, integrated by government officials into the international fur trade under the aviamento model, and used to fight and settle the Asurini and Araweté, who were also seen as threats to extractivism. Fisher (2000) notes that the Xikrin entered the fur trade to meet their own goals, not simply because of government coercion. Government-issued firearms offered security after several decades of war and migration, and long treks in search of furs allowed Xikrin families to avoid sad and dangerous reunions with the spirits of the many ancestors who died during contact and lingered in villages. In many ways, the Xikrin think of the fur boom as a golden age of peace, security, and impressive access to goods. With its own source of goods, the tribe was not vulnerable to the vagaries of government spending. However, the international ban on the fur trade generated that vulnerability and prompted the government and tribe to reorient their efforts around Brazil nut collection.

While the Brazilian state played little direct role in the rubber boom, it has entangled itself in contemporary manifestations of the aviamento system as an intermediary. By the 1970s, FUNAI had "pacified" and "villagized" many of Altamira's indigenous people and had drawn them back into extractivism through trade in fur and Brazil nuts. By mediating indigenous peoples' involvement in extractivism, FUNAI used indigenous labor to underwrite its own operating costs and established itself as a powerful supplier of manufactured goods. Extractivism turned indigenous groups into "self-financing and dependent communit[ies] . . . initiated into a system that [has] been in existence since the rubber boom in Amazonia, a system still widely practiced on Indian posts" (Fisher 2000, 77). In this context, however, FUNAI's power is largely dependent on fluctuations in the world market in nontimber forest products. Writing just as AmazonCoop was being founded, Fisher describes the Brazil nut harvest as "a calculated gamble" and "a time of great anxiety" for Xikrin villagers (2000, 78).

Extractivism creates a delicate and competitive set of alliances among Amazonian indigenous peoples, FUNAI, and corporations. Goods and services are critical forms of currency that make those alliances possible. Nonindigenous people seek raw materials such as spices, rubber, fur, timber, minerals, and Brazil nuts, and they depend on a predictable and controlled labor process to obtain these goods. Manufactured debt and direct government mandates shape indigenous people into "useful" extractivists. For their part, indigenous people seek manufactured goods and development projects that, in addition to their instrumental value, have the political value of reinforcing the power of chiefs as intermediaries with the white world and redistributors of trade goods. These material bases of chiefly power have become especially important because the dispersed and family-based nature of extractivism has weakened chiefs' traditional roles as organizers of communal labor (Fisher 2000).

Fisher's account of Kayapó-Xikrin involvement in extractivism is particularly rich because it explains the sources of indigenous people's power and agency in these negotiations and conflicts: their labor power, ecological knowledge, ability to disrupt nonindigenous economies, and ability to fall back on traditional hunting-planting-gathering livelihoods that permit relative autonomy. Indigenous people have been a resource and threat to extractive industries, and they have generally remained resilient to economic booms and busts. Their ability to shift between government and corporate partners has also been helpful; for example, a downturn in the FUNAI-organized Brazil nut trade in the late 1990s prompted tribes to fulfill their demand for goods through alliances with mining and timber firms. This is where the story of AmazonCoop begins.

Like previous extractive projects in the region, AmazonCoop grew from the convergence of foreign demand and local and international entrepreneurship. Through the cooperative, indigenous people, corporations, and the state sought to capitalize on local natural resources within the political and economic context of globalization and neoliberal policies. The cooperative's adaptations, like those of nineteenth-century seringalistas, were shaped by previous extractive projects. AmazonCoop was established in a region structured around the old aviamento system, where indigenous people and other extractivists have often been controlled by debt and violence in a nearly lawless land, and where they typically have occupied the most unfavorable position within a value chain controlled by foreigners and a large number of intermediaries. Worse, while indigenous people historically found power in their ability to provide or withhold critical knowledge, to play competing firms against each other, to violently disrupt the riverine trade of rubber and basic necessities, and to return to traditional means of subsistence, state-led "pacification" and "villagization" have significantly limited many of these avenues of indigenous power. In addition to this historical legacy, AmazonCoop also embodies several distinctive features of contemporary society, most notably the prominence of international civil society, new niche markets for socially responsible consumption, the improved legal and social status of indigenous people in Brazil and beyond, and consumer concern for the people and ecology of the rain forest.

CONTEMPORARY AND HISTORICAL EXTRACTIVISM IN THE BRAZILIAN AMAZON: ANALYZING AMAZONCOOP AS AN AGENT OF CHANGE

Cooperative organizations are a promising addition to the extractivist economic landscape in this region. Cooperatives are meant to benefit members by changing their position within production, marketing, and/or consumption systems. By drawing members together in a single institution, they can provide important benefits like economies of scale; access to credit, capital, and machinery; possibilities for providing social services en masse; and the ability of members to mobilize and exert political influence. In our previous research, we have shown that the AmazonCoop trade relationship did provide significant material benefits to indigenous extractivists in Altamira. By combining vertical integration with The Body Shop's above-market price, the cooperative was able to pay indigenous harvesters more than double the local rate for riverside purchases of Brazil nut.

However, I argue that if cooperatives are to be agents of change, they must do more than help members benefit from existing political economies; they must also change unjust patterns of power and exploitation within those political economies. To what extent did AmazonCoop change historical relations of power, control, and exploitation? On the surface, the (nonindigenous) cooperative directors seemed to be on the right track. They listed self-sufficiency, cultural integrity, and the ability to engage effectively with *o mundo branco* (the white world) as their most important goals for indigenous members. The cooperative's success in vertical integration of the rubber industry and economic diversification (results of large infusions of foreign capital) also seemed promising. Unlike the rubber tappers of the past, the cooperative was able to eliminate several intermediaries by purchasing its own cargo boat, warehouse, and oil-processing plant. This ensured the cooperative a larger profit and fewer risks than historical seringalistas enjoyed, and therefore reduced the need for extreme exploitation and violence.

On closer examination, however, AmazonCoop reproduced many historical relations of control and exploitation without enhancing the bases of indigenous people's power. The continuities in Amazonian extractivism are particularly evident in the organization of Brazil nut collection. As with previous modes of extraction, the Brazil nut trade is internationally oriented, with a market structure determined by forces outside of Brazil and its indigenous communities. The costs, risks, and difficulties of Brazil nut extraction resemble those of rubber (though they are less extreme), and the cooperative's extractive program is organized along the lines of previous FUNAI programs, which are themselves variations of the aviamento system. Laborers are controlled by the cooperative's and FUNAI's near monopoly on manufactured goods and development projects, by debt accrued through advanced supplies, and by strong government control over indigenous villages.

AmazonCoop's Brazil nuts are harvested directly from the forest, a difficult, time-consuming, and dangerous process. In most member villages, the nuts are collected in February and March by individual families, some of whom spend the entire two-month harvest season in the forest. Traveling to distant Brazil nut patches requires a significant investment of time and money for inputs such as gasoline, food, and hunting and fishing supplies. To offset these costs and encourage intensive dedication to Brazil nuts, the cooperative advances money and supplies at the start of each harvest season, which harvesters must repay in nuts. The extractivists we interviewed gave mixed reviews of this credit system, complaining about low end-of-season profits but content with the higher yields made possible by more time in the forest. Long absences threaten family agricultural production, however, and may decrease food quality for the rest of the year. One Asurini family dedicated themselves so intensively to Brazil nuts that they did not grow their own *farinha* (yuca or manioc meal), marking the first time that anyone could remember a family lacking this staple food. Though not as exploitative as the traditional aviamento system, AmazonCoop's credit system creates dependencies on the cooperative, the government, and the Brazil nut harvest by encouraging extraction at the expense of other livelihood options.

If the cooperative had been owned and operated by its indigenous members, this manner of organizing extractivism would have been a significant change from the aviamento system: indigenous harvesters would suddenly have controlled the capital, credit, and equipment that shaped their work, and their production of Brazil nut oil would have moved them up the value chain. Unfortunately, however, AmazonCoop was a top-heavy organization in which power and decision-making authority were concentrated not only outside of indigenous villages but also outside of Brazil. The cooperative's use of "them" in public references to indigenous members is revealing in this regard. AmazonCoop was an organization of nonindigenous advocates acting *on behalf of* the indigenous people of the Altamira region and in the interest of a multinational corporation and a government agency. The cooperative's strategy was devised by an international advisory board composed of representatives from multinational corporations and leaders of the corporate social responsibility movement. This strategy was then handed down to an administrative council whose active members were all nonindigenous Brazilians. It was only through village representatives to the General Assembly that indigenous people were involved in decision making, but assemblies were rarely held, and a quorum required the presence of as few as three indigenous people. In addition, the cooperative's three main income sources—Brazil nut sales, the provision of Internet services, and donations via the Tataquara Eco-Lodge—all depended on nonindigenous managers and foreign money and markets. Indigenous people had virtually no say in the organization of these economic projects and controlled none of the assets on which they depended.

AmazonCoop probably did not choose this international orientation for the purpose of maintaining the dependency of indigenous members. In addition to extremely low levels of education and professional capacity in Altamira's indigenous villages, there were several good reasons that the cooperative founders might have looked abroad for support. The cooperative was founded by nonindigenous entities—The Body Shop and the regional FUNAI administration—to meet the material, financial, and marketing needs of those entities. In addition, this strategy was an economically rational response to the cooperative's ability to mobilize significant levels of foreign capital from countries where indigenous Amazonians have high symbolic power and to the difficulties of finding domestic support as a result of anti-indigenous racism. Finally, the cooperative emphasized international partnerships because, in the view of its managers, the Brazilian state had failed to provide adequate support for indigenous people. Nonetheless, while international links facilitated access to new markets, funding sources, and expertise, they did little to eliminate the dependencies and power imbalances of previous extractive systems.

In fact, the cooperative decreased many of the traditional sources of indigenous power within extractivist economies. By integrating indigenous people, the state, and corporate interests into a single institution, it reduced competition between corporations and the state and limited opportunities for indigenous groups to play outsiders against each other. That it eliminated competition without securing a long-term commitment or formal contract from The Body Shop made the economic benefits of this "fair trade" system even more precarious. In addition, because indigenous labor contributed to less than half of the cooperative's total income (the Internet service provider was equally lucrative), their ability to use slow-downs and strikes to pressure Brazilian and international decision makers was limited. The extent of government control over indigenous villages, development and social service opportunities, and the cooperative further enforced indigenous people's compliance with the cooperative's labor demands.

The final days of the cooperative revealed the full extent of indigenous disempowerment. In the dramatic conflicts that led to the cooperative's dissolution, the organization's enormous debts and legal liabilities were transferred to a newly appointed indigenous president while a sizeable portion of machinery and other assets mysteriously disappeared or were transferred to nonindigenous people. Indigenous people not only lacked effective decision-making power, even the collective nature of cooperative economic resources was a farce. Given this, the cooperative can best be seen as a new intermediary in an aviamento-like system. Rather than enabling indigenous people to control their own capital, credit, and labor, the cooperative maintained indigenous people at the lowest rungs of the production chain, albeit with better compensation.

Of course, better compensation and decreased exploitation are important gains, and the indigenous villagers that we spoke with were clearly happy with the higher incomes that the cooperative permitted. My point is simply that these benefits were secured through the workings of a highly unequal society and did not entail a change in indigenous people's position within local power relations.

Furthermore, many of these benefits depended more on external factors than on the cooperative itself. For example, because Brazil nuts have a shorter harvest season and constitute a much smaller industry than rubber, harvesters are able to pursue other subsistence activities and are less likely to become dependent on this single commodity and the intermediaries who control its sale. Also, the security of a "pacified," government-monitored indigenous labor pool lowers economic risks and the need for more extreme forms of labor control. Most important, though, the main value of the cooperative's Brazil nuts derives not from their use-value but from their symbolism, from the positive image that comes with fair trade, cooperative organization, and indigenous development. This fact alters the industry, establishing a limit to the exploitability of indigenous people and obliging corporate and state partners to provide (or at least appear to provide) something recognizable as "development."

CONSIDERING POWER AND HISTORY IN COOPERATIVE DESIGN

By now it should be clear that AmazonCoop provided economic benefits to indigenous people in the Altamira region, but it did not change (or even attempt to change) their extreme marginalization within networks of power. Unfortunately, this is not an unusual outcome of cooperative organizing. The Amazon-Coop story can be useful, however, for reimagining how we might transform cooperatives from beneficial social and economic institutions into agents of profound social change.

The key, I think, lies in the way that we think about cooperative design. If we focus on developing cooperatives that can successfully execute economic projects, that is precisely what we will get: cooperatives that execute economic projects and marginally improve people's well-being. This work is important, but these cooperatives will tend to work within—and therefore reproduce—the power dynamics that have created today's vastly unequal world. What if we design cooperatives based on a different set of questions? What if, instead of (or alongside of) the standard question of how to organize economically viable cooperatives, we also thoroughly interrogate how our cooperatives can challenge the historical relations of power that have created the very need for institutionalized cooperation among the poor?

AmazonCoop provides a useful example for how cooperatives might be designed differently. The founders of AmazonCoop asked, "What type of organization do we need to create in order to trade with those multinational corporations that will pay a premium in order to appear fair and just?" This question is perfectly reasonable, but it reduces indigenous people's interests to the narrow issue of achieving a higher price for their products. It also puts multinational corporations in the driver's seat by emphasizing the need to adapt to their demands. Most important, it diverts attention away from critical reflection on the status quo. Instead, we might place this new possibility of cooperative-based fair trade into the historical context of indigenous involvement in extractivism, asking how it could alter the resources, challenges, and vulnerabilities that shape indigenous people's social and economic opportunities. While there are many ways to approach this discussion, I offer the following questions as one path for considering how historical power relations can be mobilized and changed in order to promote indigenous well-being today:

1. How can indigenous people use this cooperative to preserve and expand historical sources of power in extractive economies?
2. How can indigenous people use this cooperative to mobilize new types of power available in today's global economic and social landscape?
3. How can indigenous people use this cooperative to avoid, escape, or change the dynamics that have contributed to their unfavorable position in extractive economies?

These questions clearly establish that the cooperative is not an end in itself but rather a tool that indigenous people might use to combat their systematic marginalization. By making power central to cooperative design, they help us look beyond the ways that indigenous people can use the cooperative to make inroads with a multinational corporation willing to pay a premium for the right to use indigenous, environmental, and fair trade symbolism; they also prompt us to think about how indigenous people can carve out power within that trade relationship and how they can leverage that relationship to make broader claims to power.

Examining the old aviamento system, it is clear that a lack of capital maintained indigenous people in the lowest links of the commodity chain. A genuinely participatory cooperative would create the possibility of indigenous-controlled credit to support the Brazil nut harvest and indigenous-owned transportation and processing plants. These new commons would surely improve the position of indigenous people within the commodity chain, but the series of questions that I propose raise other possibilities as well.

For example, history shows that indigenous people sometimes resorted to extreme measures—disrupting trade through violent river closures or leaving the rubber trade altogether—in order to force consideration of their interests. This historical understanding could be used to ensure that any new cooperative actively privileges indigenous interests. First, we might institutionalize frequent and meaningful opportunities for indigenous people to express their interests in ways that do not disrupt trade, for example through regular meetings on indigenous lands. But a cooperative that seeks to actively challenge power hierarchies might take a second, more unusual step of valuing and even strengthening indigenous people's ability to disrupt trade. This type of ultimate veto power—which could be established through a number of different mechanisms—subordinates the cooperative economic project to the broader goal of social change. The cooperative might also strengthen indigenous power vis-à-vis international partners and other local economic elite (including the government) by using profits to strengthen traditional subsistence systems and to pursue new livelihood options, thus decreasing dependency on individual industries and partners. Not incidentally, designing these possibilities for protest into the cooperative might actually stabilize (rather than threaten) international partnerships by decreasing the risk of unexpected and more dramatic forms of resistance. If indigenous people have the power to demand meaningful dialogue, they don't need to assert their power through mass mobilizations, protests, and media attacks against their partner corporation.

This brings us to a second point: how the cooperative might use contemporary developments to promote change. AmazonCoop was founded on the new symbolic value of indigenous people and the interest of international civil society in indigenous rights and conservation, but these resources were used only to increase profits to The Body Shop and earnings for indigenous harvesters. The lack of engagement with indigenous political movements is especially surprising given that the initial idea for The Body Shop's involvement in the region grew out of a masterfully executed symbolic protest by local indigenous people against the Belo Monte hydroelectric dam. A lucrative project like the AmazonCoop Brazil nut trade might lend extra weight to indigenous demands that the Brazilian government uphold basic rights, particularly in areas like land rights and education, which become (monetarily) valuable prerequisites for foreign direct investment. This type of project could also be used to unite tribes that were previously divided (and exploited through their division) and to form mutually beneficial relationships between urban and rural indigenous people.

These ideas offer a hint of the possibilities that emerge when we consider cooperatives as agents of change rather than more limited economic projects, and when we place them in their historical context. Of course, most cooperatives will not be able to launch entirely new modes of production, extraction, exchange, and distribution. Historical relations of exploitation, dependency, and unequal exchange may be so entrenched that cooperative members cannot create a radically new alternative all at once. In fact, some of that resistance will come from cooperators themselves; developing an intertribal cooperative in the Altamira region would challenge traditional systems of chiefly power and authority, and parallel projects in other rural areas would provoke similar challenges. But these sticking points only underscore the importance of more transformative cooperatives. They raise a final question that could aid in rethinking cooperatives:

4. How can cooperative members engage in these still imperfect, still exploitative economic practices in a way that promotes their long-term liberation from systems of exploitation?

Change will not come overnight, neither in cooperatives nor in society at large. Many new AmazonCoops will be created, projects that don't challenge the status quo but do improve living conditions for the economically disenfranchised by offering them slightly higher profits and added social perks. These remain valuable. But I think that cooperatives can aspire to more. Through critical interrogations of power and history, I think cooperatives can help us imagine and move toward a more profound reconfiguration of power in the future. This is surely a worthwhile task.

COOPERATIVE EXPERIENCES IN COLOMBIA

SMALL PRODUCERS AND COOPERATIVISM IN COLOMBIA

MIGUEL RICARDO DÁVILA LADRÓN DE GUEVARA

TO APPRECIATE SMALLHOLDER COOPERATIVES and associations in Colombia, it is important to understand the contemporary characteristics of the rural economy, its importance nationally, and the major threats it faces. In this chapter, I present a vision of small producers in Colombia and the role that some significant cooperatives have played in small-farmer development. This chapter is based on the results of rural-sector studies carried out since 1980 by the School of Rural and Environmental Studies at the Pontificia Universidad Javeriana. The first part of the chapter focuses on the smallholder economy. The analysis draws primarily upon the work of Jaime Forero Álvarez (2010), a professor in the Department of Regional and Rural Development who has addressed the profound changes incurred by the smallholder or family economy throughout the twentieth century and the challenges faced in the last 25 years—including violence, land seizures, displacement, and, as a result, massive impoverishment. In the second section, I lay out the importance of the cooperative sector in supporting the smallholder economy, grounding my discussion in studies carried out by the Solidarity Studies Unit (Unidad de Estudios Solidarios, or UNES) in the southern provinces of the department of Santander (see López in this volume).

THE PEASANT ECONOMY IN COLOMBIA

The family-based small-producer economy holds an important place within Colombia's agricultural sector. According to Garay, Barberi, and Cardona (2010), 87 percent of productive farm and livestock units in Colombia are associated with the smallholder economy. From data collected for a 2005 household survey by the National Administrative Department of Statistics (DANE), this is equivalent to about 1.4 million households (almost 6 million people, or 12 percent of the total number of households in the country)—without including the 760,403 family groups who were displaced between 1998 and 2008 (Forero 2010, 74). On average, the productive agriculture and livestock units linked to the smallholder economy use 4.8 hectares of land and allocate at least part of their labor and income to their farms or plots. Such numbers contrast with those pertaining to business agriculture, which uses an average of 65.1 hectares per farm, with 53.1 hectares dedicated to pastureland for grazing livestock (Garay, Barberi, and Cardona 2010).

Beyond doubt, the contribution of family production to national livestock and agricultural production is of great importance. Several studies carried out over different historical periods clearly illustrate this. For example, in 1988, the Agriculture and Livestock Sector Studies Mission (Misión de Estudios del Sector Agropecuario, or MIESA) showed that smallholder agriculture in Colombia constituted 57.1 percent of crops and contributed 42.7 percent of agricultural production value—this, without counting coffee, sugar, or livestock. Forero's calculation more than 10 years later separated small-scale from capitalist agricultural production, averaging for 1999–2000, and found that predominantly smallholder production made up 67.3 percent of harvested land and 62.9 percent of national agricultural production value without counting coca or poppy farming in either case. Another study, by Garay, Barberi, and Cardona (2010), using data from the National Households Survey, calculated the smallholder economy as 47 percent of total land harvested in 2005 and 50 percent of production value for transitory crops; in the case of permanent crops, this participation reached 56 percent of harvested area and 48 percent of production value.*

*Forero notes that these data are not comparable to those that he calculated for 1999–2000 because they come from different sources.

In the case of coffee, smallholder production has played an increasingly important role. Today, this sector occupies 78 percent of land planted with coffee crops. Moreover, the small-producer sector's great capacity for adaptation as a result of a combination of monetary and nonmonetary strategies has given it greater resilience in comparison with industrial (or capitalist) production (Forero 2010, 72). And as for livestock production, small producers hold 69 percent of land dedicated to ranching and own 14 percent of cattle at the national level (Ibid., 70).

TRANSFORMATION OF SMALLHOLDER HOUSEHOLDS IN COLOMBIA

One of the most significant changes in the Colombian smallholder economy has been the partial monetization of family agricultural production systems and smallholder subsistence. This change is the result of smallholders' increased entrance into labor markets (both supply and demand), agricultural inputs, goods, and services. This is different from what was happening in the mid-1900s, when small-producer households accessed agricultural inputs and other consumer goods through their own or their neighbors' farms and used mainly family labor. Today, the family-based producer sector in Colombia hires between 17 and 53 percent of the labor employed on its land. Furthermore, this sector purchases "a high percentage of foodstuffs, [and] almost all their clothing, housing materials, vehicular transportation, public services, health, education, electrical appliances, beverages," and so on, on the market (Forero 2010, 80–83).

Greater dependence upon consumer goods and monetization has helped transform economic rationality within the smallholder sector. To wit, we have seen the development of a particular production logic involving two clear spheres. In the monetary sphere, producer families carry out their market transactions to acquire goods and services. And, in the domestic sphere, producers use their own farm resources, as well as family and outside labor, according to relationships of donation and reciprocity (Ibid., 85).

Another area of generalized transformation has been in technological change, which, according to studies evaluating Colombia's Integrated Rural Development program (DRI, in Spanish), has been evident since the 1980s. Conclusions from a study carried out by MIESA in 1990 showed that this change, based on the introduction of Green Revolution practices, narrowed the technological gap between small farmers and big producers (Ibid.). Such evidence, in turn, belies assertions made even today about the resistance of rural producers to technological and technical change.

An additional important development in the family agricultural sector has been members' participation in marketing their products at different levels. This sector allocates a good portion of its harvests to the market through either informal intermediaries or supply warehouses in big cities and market squares. As Forero explains, the informal and formal marketing channels complement one another when it comes to satisfying low-, median-, and high-income consumer needs (Ibid., 91).

Higher participation of small-producer households in the market economy does not mean that local consumption is no longer important, however. As Forero highlights, local consumption among small-producer households is on average around 30 percent. Indeed, there are novel practices in sales and product placement whereby certain producers who achieve high levels of local consumption manage to significantly expand their market sales thanks to the complementarity between the domestic and monetary spheres of their production systems.

It is also worth pointing out a particular change in the conceptualization of microscale economies. In Forero's research, for example, case studies demonstrate that small producers have achieved profits similar to or higher than those of medium and large producers despite attendant costs, including remunerating their own labor force. Without generalizing, what is striking is the assertion that "in some cases, it can be shown that medium [producers] do not have the advantages belonging to large-scale production, nor do they benefit from certain possibilities that small producers have to save costs" (Ibid., 93). For Forero, small producers' focus on tasks specific to their farms generates microeconomies of scale derived from "the virtuosity of the farmer" in attending directly to his farm and to the work it demands. This gives the family farmer a comparative advantage vis-à-vis large-scale farmers and/or agriculture and livestock businesspeople.

Finally, Forero responds to stark affirmations related to family producers' low income and high poverty levels, which have served arguments that smallholder production is unviable. Based on studies carried out since 1990 in diverse rural areas of the country, he concludes that, "as far as our analysis goes, family production is economically viable if we take labor remuneration, precisely, as an indicator of viability against other alternatives. But what is perhaps most notable is that, even after paying expenses, the producers analyzed offer payments that remain higher than the minimum wage" (Ibid., 97).

THE SMALLHOLDER ECONOMY IN THE AGRARIAN CONTEXT OF COLOMBIA

The Colombian small-producer population, with its multiplicity of productive means, forms part of an agrarian structure that interconnects them with other economic agents. Forero highlights seven types of agrarian spaces, in five of which the smallholder economy has a significant presence. These spaces are: (1) spaces of "rural capitalism," wherein family enterprise coexists and has a working relationship with capitalist enterprise, and in which investors—who often come from smallholder communities themselves—have made possible the productive expansion and intensification of crops such as coffee; (2) spaces in which large, medium, and small producers converge through mechanized farming, wherein family production persists in areas economically dominated by capitalist agribusiness but interbusiness relations among the different types of producers are lacking; (3) small-producer spaces, wherein family agriculture predominates but rural plots and communities are "surrounded" by cattle farming *latifundios*; (4) disputed spaces between owners of latifundios and small-producer communities, wherein the latifundios have expanded at the expense of smallholder properties and as a result of rising land purchases by drug traffickers;[†] and (5) resettlement spaces, in which already-settled small producers and businesses (extractive or cattle grazing) converge with new latifundios and indigenous communities (Ibid., 74–77). Overall, it is important to emphasize that illegal crops (coca, poppies, and marijuana) occupy fewer than 200,000 hectares overall, compared to the 40 million hectares of grassland needed to accommodate a herd of nearly 30 million grazing livestock animals. Nonetheless, the coca-producing regions are characterized by the emergence of armed actors, creating a violent and unstable context for producers.

[†]These spaces are concentrated in Magdalena Medio and on the Atlantic coast.

THE DISPLACED POPULATION AND THREATS TO THE SMALLHOLDER ECONOMY

Over the last 25 years, the rise in paramilitary action, drug trafficking, and guerrilla warfare, together with the push toward neoliberalism by the César Gaviria government (1990–1994), all contributed to the dramatic impoverishment of the 760,403 family groups who were displaced between 1998 and 2008 (Comisión 2009). Not only stripped of their land, they were dispossessed of their crops and animals, as well.‡ It is worth stressing that the total area abandoned between 1998 and 2008 is equivalent to 25 percent of farming land in the country. These families' basic capacity for generating monetary and nonmonetary incomes was substantially affected. If 49 percent of these families were classified as not poor prior to displacement, today only 3.4 percent qualify as such; and while predisplacement 31.5 percent were indigent, today 80.7 percent are (Forero 2010, 102–103).

Additionally, the increasing concentration of land in the hands of latifundio owners who dedicate it to cattle grazing has contributed to displacement and war in the countryside. In a total area of 42 million hectares, 32 million are dedicated to cattle grazing latifundios, 6 million to semi-intensive cattle farming, 1.6 to industrial agriculture, and 2.6 to family agriculture. This inequality in land distribution allows Forero to highlight the following paradox: "While the productive capacity of family farmers cannot be minimally activated due to limitations on access to land, most of this resource [land] finds itself monopolized in an unproductive state by latifundios dedicated to cattle grazing" (Ibid., 105).

COOPERATIVES AND THE PEASANT ECONOMY

Cooperatives appeared in Colombia in 1931 after the issuing of Law 134, and their presence was felt mainly in urban sectors. In the rural sector, cooperatives appeared in the 1940s and began to have an important presence in the 1960s with the enthusiasm inspired by the Cuban revolution and as an instrument

‡According to data in the Second National Survey on the Verification of Rights of the Displaced Population, 55.5 percent of displaced persons owned land, and 78.9 percent owned animals that generated income (see Forero 2010, 102).

for modernizing the countryside. From then on, the state pushed mainly agrarian reform cooperatives, the Catholic Church promoted rural savings and loan cooperatives, and dairy businesspeople, along with the sugarcane and coffee unions, promoted agro-industrial and marketing cooperatives. These experiences were consolidated starting in the 1980s. Currently, cooperatives form part of the Solidarity Economy System, which in Article 2 of Colombia's Law 454 of 1998 is designated as the "socioeconomic, cultural, and environmental system shaped by a collection of social forces organized in associative ways identified by solidarity, democratic, and humanist self-governing practices, not motivated by profit, for the comprehensive development of the human being as a subject, actor, and ultimate beneficiary of the economy."

The most important results achieved by cooperatives in support of the smallholder economy can be organized into three fundamental categories: those of an economic character, those that concern social capital, and those related to technological change. These are explored below with reference to studies in the south of the department of Santander. Five cases are discussed in detail by López later in this volume.

RESULTS OF AN ECONOMIC NATURE

Cooperative experiences in the southern provinces of Santander exemplify the role that cooperativism has played on behalf of small producers, insofar as it has allowed them access to financial and other complementary services. This has unfolded within the framework of the Catholic Church's rural development programs, thanks to which rural savings and loan cooperatives (*cooperativas rurales de ahorro y crédito*, or CRAC) have become active agents for local and rural development. In this vein, the Secretariat of Social Ministry (Secretariado de Pastoral Social, or SEPAS) of the San Gil and Socorro diocese started a program in 1964 that "has contributed to strengthening the socioeconomic fabric of this peripheral region" in large part because SEPAS knew how to "extract the best from the cooperative model and from its adult education methodologies to develop and redevelop original development models and strategies adapted to its particular context" (Bucheli 2006, 13).

By 2007, this network of cooperatives supported the existence of seventeen CRACs that managed to consolidate autonomous capital from member contributions. This allowed them to cover unsatisfied demand for credit through the successful actions of loan operators to express social and economic rationality in local cultural terms (Dávila 2004; Silva and Dávila 2002). In 2005, 15 of these CRACs provided loans of close to $31 million (López and Peña 2005).

Despite these achievements, a weakness remains in the fact that the most needy sectors have not been totally addressed. Within areas under the influence of CRACs, there remain landless producers who, in their vast majority, are not associated with cooperative practices. They have neither sufficient money nor assets to allow them to join and use the savings and loan services that these organizations offer (Dávila 2008).

What has been most important in this process has been that the CRACs have managed to carry out activities that are well established and well connected to local and regional communities. This success is due to two strengths: "For one, the close relationships with members, which reduces the problems of poor selection and moral risk. This facilitates loan repayment and allows the institution to tailor itself to borrowers' needs. And second, financial administration by directors and managers that is derived from knowledge of microcredit" (López and Peña 2005, 47–48). In addition, the offer of "tailor-made services for the needs of members" and "guarantees founded in trust and familiarity" makes room for flexible and timely lending services (Dávila 2008, 312).

RESULTS RELATED TO THE CREATION OF SOCIAL CAPITAL

Lobo (2001) shows how a cooperative in the south of Santander has been an important actor—together with the parish and the agricultural school—in the peace process. Facing pressure from an armed group that had been dominant since the mid-1980s, this area decided to become a "peace territory" in 1997. Given the absence of the state and the presence of armed actors, the community decided to organize itself into committees and create a space to resolve conflicts through dialogue, as well as address specific community needs and outline demands for recognition as part of civil society, including respect for human rights. In this way, they pushed the pacification process forward. The contribution of the COAGRANJA cooperative was to strengthen community identity through a process of "building values, attitudes, [and] social norms based principally on trust and solidarity (cognitive social capital) from the time of its founding by offering loan services based upon personal guarantees based on trust and familiarity among members (the use of a guarantor)" (Ibid., 103). They also facilitated the presence of members and employees as cooperative representatives at different moments of the pacification process, such as meetings with the armed groups and committees for human rights.

RESULTS RELATED TO TECHNOLOGICAL CHANGE

The CRACs created in this territory soon moved beyond the distinct stages of their life cycles (i.e., promotion, creation, development, and consolidation) and strengthened their economic positions. Meanwhile, complementary services related to farm and livestock activity were converted into basic support for improving technical conditions of production and adopting new technologies appropriate to members' economic circumstances. For example, the cooperatives located in Barichara (founded in 1965) and Villanueva (founded in 1960), with significant activity in bean crops, established tractor lending services in the 1980s—at fair prices—facilitating members' work on the flatlands of the two municipalities. By so doing, they accelerated the adoption of modern farming technology in the subregion. Likewise, the cooperatives were able to offer rentals of grain strippers and other necessary tools at reasonable prices to introduce modern technology into different farming tasks (Medina 2002).

The La Granja cooperative (founded in 1974), located in the municipality of Sucre, stimulated a comprehensive development project starting in 1989. Supported by the Inter-American Foundation, which promoted technical assistance services, this allowed for increased milk production and improved cheese production. The Valle de San José cooperative (founded in 1967) promoted an institution-building process in the 2000s to improve members' agriculture and livestock production, pushing for environmental education in agroecology and technical assistance for organic agriculture and solid waste management. This effort has made it into one of the municipalities with the highest numbers of production systems aimed at clean farming (Ariza 2007).

CONCLUSIONS

The most important conclusion derived from the role of cooperatives in the context of the smallholder economy has to do with the skills these cooperatives have developed in approaching local culture—especially in interpreting the particular production logic held by family producers—as they offer their different services. The ability to consolidate autonomous capital from member contributions, for instance, has provided flexibility in the terms set for loan repayment. Also, close relationships with and between members allows for decisions regarding credit to be based on notions of trust and awareness of household's needs, ability to repay loans, and credibility. The role of cooperatives in strengthening community identity in a violent context and rebuilding trust and solidarity has been particularly important in the Colombian context, creating real possibilities for a viable future for smallholders. In other words, cooperatives have advanced the sustainability of the smallholder economy and its technological development while generating ideal conditions for its maintenance and defense within a violent setting. Through the development of social capital, they have also helped support and promote the peace processes that have taken place in the region. In this sense, a very important contribution has been their advocacy for small producers as citizens, as subjects with rights and responsibilities, and in empowering and defending small producers as social subjects worthy of full recognition by the rest of society.

COFFEE ASSOCIATIONS IN COLOMBIA

Strategic Changes for Adapting to the Global Context

LUIS ALBERTO CUÉLLAR GÓMEZ AND
MARÍA ISABEL RAMÍREZ ANAYA

IN 1989, THE RESTRUCTURING OF THE GLOBAL ECONOMY through open markets, free trade, and common market blocks took a toll on the coffee industry by forcing the dismantling of the International Coffee Agreement. The International Coffee Organization had handled this agreement since the 1960s, managing to stabilize international prices by setting annual export quotas for each producing country. This, in turn, facilitated the overall development of the coffee sector, as well as long-term economic planning by large producing countries, such as Colombia. Colombia itself specialized in exporting green coffee as an excellent-quality raw material to be mixed with beans from other origins, a strategy that allowed for the growth of coffee-roasting facilities, which entered into supply quota agreements between exporters and the National Federation of Coffee Growers of Colombia (FNC, in Spanish).*

Despite the fact that over the course of the twentieth century Colombia irrefutably created a comparative advantage for its coffee industry in terms of domestic infrastructure, crops, research, technology, and internal marketing, producers failed to acquire external marketing experience. On the one hand,

---

*The FNC is internationally recognized by the famous image of Juan Valdez. The FNC, since its formation in 1927, has represented over 500,000 coffee-growing families—the majority of which are small producers—in promoting the production and export of Colombian coffee.

specialization was emphasized and supported through monoculture and the intensive use of external agricultural inputs under the technological model of the Green Revolution. This emphasis on modernizing peasant agriculture came with high costs, including the rapid degradation of natural resources, the gradual loss of local technical knowledge and crop diversification as a development strategy, and the acceleration of migration from the countryside to the cities, as small farmers did not have the resources to follow the modernization trend, being forced, instead, to abandon production altogether. In addition to these, the importation of food has tended to further destroy the productive foundation of local agriculture.

On the other hand, the oligopolistic structure built around the International Coffee Agreement and its quota system historically kept small coffee producers in the role of dried parchment coffee providers for internal and external marketers. In this way, they were denied the possibility of building a product with value added due to its distinguishing features such as the type of crop, the altitude, the varieties planted, and the collective intellectual property protection furnished by the terroir denomination (based on qualities specific to produce from a given region).

Today, all of this has changed radically. While it is true that the opening up of the international coffee market negatively impacted the large coffee-producing countries, it also allowed for the development of a market for sustainable, organic, certified, and specialty coffees. This market emerged in the mid-1990s, backed by demand from consumers willing to pay more for coffee produced according to universal standards that met protocols for social justice, environmental protection, and the implementation of best practices in agriculture and manufacturing. Although these standards have benefited small producers by making them the preferred source for high-demand coffees, the standards themselves are established by certification bodies in consumer countries in Europe and North America. Generally speaking, these bodies do not consider the local conditions and complexities of the producing regions.

In this chapter, we draw from several years of work as practitioners with ACDI/VOCA's Specialty Coffee Program, implemented in five coffee-producing departments (or states) of Colombia from 2002 to 2010 with funding from USAID. The program was specifically designed under the framework of an alternative development strategy to provide small farmers a way out of coca production. Here we examine some of the local conditions and complexities that have made small-coffee-producer associations and cooperatives key entities in negotiating the vast transnational terrain of specialty coffee markets. To wit, small-producer associations and cooperatives have worked to transform smallholders' place within international trade networks by connecting producers, through direct market relations, with roasters and importers at the other end of the value chain. They have also worked toward strengthening producers both economically and socially, with institutional backing, to help them transition away from the production of illicit drugs. The impacts have been transformational in terms of improving quality of life and empowering the households and communities where producer associations and cooperatives are located. This is especially important in areas that are transitioning away from coca production in order to minimize the presence of armed actors (see Dávila, this volume).

We begin by identifying the challenges small-producer associations face when attempting to integrate into the value chain and the global marketplace, and we describe the strategies that, in our experience, have been most successful for addressing these challenges. We examine producer organization experiences in the Huila region, focusing on a specialty coffee support program led by ACDI/VOCA. We wrap up our discussion with a case study that exemplifies the changes taking place among producer organizations adapting to the specialty coffee market.

THE CHALLENGES AND DIFFICULTIES OF GLOBALIZATION

In the face of economic globalization, Colombian coffee growers and supporters have had to seek locally specific economic strategies. Those of us working in Colombia's coffee-growing region have had to define our future using regional and/or local value-added strategies. These strategies require an expansion of the socioeconomic dimensions of coffee growing, going beyond the agricultural activity of farms to encompass other related activities such as marketing, industrialization, and training. They also require—above all, in the case of the small coffee producer—that people work in a coordinated manner by forming different types of producer organizations (OPs, in Spanish). OPs have spread rapidly and enjoyed notable success on various fronts, including market presence, access to public goods, rights of representation, and better quality of life for smallholder households.

Value chains and global market forces create continual new challenges for small OPs, including the following: rapid acquisition of new technologies and technical knowledge in order to improve production, processing, and marketing operations; obtaining resources for joint finance of their operations; the need for highly skilled management; and achieving efficient market insertion despite strong competition from big private companies, as well as from medium and large agribusinesses. Such difficulties were clearly visible, for example, in the World Bank's 2007 report on agriculture and development, which identified key challenges for organizations and governments wanting to promote the integration of small producers and their OPs into value chains in such a way that benefited the rural economy in developing countries. Another key challenge in the Colombian case is the presence of armed actors in the rural context, where farmers feel forced to produce illegal crops, partly because of lack of viable economic alternatives but also as a result of coercion and intimidation.

Through ACDI/VOCA, we worked to support peasant coffee farmers as they transitioned from producing low-value commodity coffee and, in some cases, illegal crops, to high-value specialty coffees. Over the course of more than five years, our work included technical assistance for farmers, the introduction of new processing technologies, work with farm laborers, broad-scale training in coffee quality and differentiation, and ultimately the establishment of trade relationships between OPs and buyers. This experience with small-coffee-grower associations in Colombia and other parts of Latin America highlighted the importance of the above-mentioned challenges in the development of cooperatives and associations, particularly in a context where small producers have historically been excluded.

FINDING A BALANCE BETWEEN SOLIDARITY AND ECONOMIC EFFICIENCY

A common dilemma facing small OPs is the need to find and maintain a balance between economic efficiency and social equity—two objectives that tend to generate conflict since they often represent opposing forces. Operating in rural communities that place a high value on social inclusion and solidarity, small OPs find it difficult to exclude association members who do not fulfill their obligations. In their attempts to protect and maintain solidarity, organizations are pressured to subsidize the least productive members at the expense of the most productive members. Conversely, they feel compelled to provide assistance and services to their communities, affecting not only their economic efficiency and innovation, but also depleting their own resources, generated through their commercial and/or financial operations.

Although small-producer cooperatives and associations should practice values of solidarity and mutual support, they need to use their market involvement as a foundation for social solidarity. That is, to be able to lend services to members and communities, they need to generate the resources necessary to do so. At the same time, services should be lent equitably, which means that there should be an awareness of social equity, allowing for the just distribution of services offered by the cooperative. Insofar as members successfully balance revenue and operational costs, the cooperative will grow stronger through the economic gains generated, adopt a more favorable sector position, and exercise greater influence on its surroundings. To these ends, and following market quality standards, all members should adopt shared policies for production efficiency and product delivery to fulfill contracts and generate hoped-for profits that will improve both the income of each associate and the financial position of the organization.

As we observed among coffee-grower associations in Colombia, subsidies and support to members who were inefficient and disloyal usually resulted in weak organizations that ultimately dissolved. To rectify this situation, solidarity and mutual support for the most needy members was channeled through technical assistance and improved coffee-processing infrastructure. In exchange, members had to be punctual in meeting production requirements, productivity and quality standards, and credit payments to organizational rotating funds, if available. It was also important to make use of purging mechanisms, which excluded members who did not fulfill their commitments and duties. In other words, while members claimed their rights, organizations demanded compliance and created exclusion mechanisms in order to avoid situations of great inefficiency.

Successful organizations will forever contend with maintaining equilibrium between social and economic interests. If a cooperative or association is managed through criteria focused mainly on solidarity and mutual support, it will suffer an imbalance and risk collapse. The opposite is equally true, insofar as complaints and associate withdrawals due to sidelining of social objectives and community dissatisfaction can lead to organizational breakdown.

IMPROVING ORGANIZATIONAL GOVERNANCE

A great challenge in all cooperative associations is how to achieve fair representation through attention to members' wide gamut of interests. Organizations should ensure that small producers' interests are represented, including among them those of women and youth; thus it is important to consider the household as the production unit. This means considering different needs in order to later address them. In this sense, the most successful coffee-grower associations and cooperatives in Colombia, in addition to promoting cooperative education processes, have promoted the grassroots participation of their members by forming various committees, each with its corresponding norms and each contributing to the management's administration of activities within the cooperative.[†] These committees pay attention to various phases and types of agricultural production, processing, and marketing, including those where young adults, men, and women can develop specific leadership skills, such as in the processing, tasting, and marketing of specialty coffees. In addition, there has been attention paid to areas where women have a predominant role, providing credit for the establishment of subsistence gardens and small animal raising.

A key program to improve family dynamics and inclusion that impacts organizational governance within the OPs is the provision of psychological assistance to help farmers deal with the stress related to illegal drug production, which has resulted in significant violence experienced by communities and producers and which eventually impacts the incidence of violence within households.

Fortifying members' organizational capacities, both as members of the OP but also as members of households, strengthens their representational voices, along with the OP's decision-making mechanisms, which ultimately contributes to improved governance. It is also important to maintain efficient information

[†]A similar strategy has worked well in Peru, where coffee-grower cooperatives have set up grassroots committees in the rural areas. These committees serve as nuclei for communication and information exchange with cooperative headquarters. This also facilitates the development of organizational leaders and promoters through the exercise of local governance; active participation arises out of discussions over technical issues, production, quality, harvesting and financing, among other matters.

and communication systems using media such as radio, technical bulletins, and balance sheets, as well as appropriate information and entertainment technologies (including virtual and/or blended learning model ones) to offer younger or newer members tools to make the most of their learning. Such measures allow organizations to improve governance and prepare leaders to fulfill their responsibilities with transparency on behalf of the social base they represent.

DEVELOPING MANAGERIAL CAPACITIES FOR VALUE CHAINS

One indispensable requirement today has to do with the development of management capacities for high-value chains, a task that Colombian coffee growers previously delegated to exporters and the FNC. Managers and their organizations have to negotiate within increasingly sophisticated national and international supply chains with strict standards and changing market situations and certifications; such is the case with prices on the international coffee market. Managers need to organize associates' inputs and production to satisfy value-chain demands while also keeping in mind the following: (a) achieving economies of scale and on-time deliveries; (b) meeting economic, social, and environmental standards required by the certification schemes that individual buyers request; and (c) complying with exporters' specifications, as well as with those of sophisticated specialty buyers or specific market niches. This means participating in analyses, dialogues, and negotiations with other actors along the chain, thereby creating direct communication links between producers and buyers.

Colombian coffee-grower associations met the challenge of increasingly complex professional management in two different ways. On the one hand, traditional coffee-grower cooperatives supported by the wider coffee-producing institutional structure, the FNC, had to restructure and modernize processes within their business culture. To adapt to new market realities, they implemented professionalization and training measures for their managers and leaders, as well as merger and incorporation processes to survive using economies of scale. On the other hand, newer coffee-grower associations, formed since the 1990s, have learned as they go about better management and business practices, adapting quickly to meet scenarios presented by the national and international coffee market. Those who have managed to do this successfully remain, but others do not. Currently, OP learning goals focus on understanding topics such as risk management in coffee, financial operations for marketing, and evaluation standards for specialty coffees. Across both sectors, the use of transparent financial management systems has been important, as have periodic presentations of results to associates and OP-served communities.

Managers of small OPs, who participate in high-level technical discussions like those surrounding local, regional, national, and/or international commercial negotiations, need to develop technical and commercial skills and tools. In some cases, they must look to outside experts and facilitators, seeking professional advice to better understand the consequences of required changes and gaining experience in preparing their input for dialogues that affect or concern them.

FACING THE EXTERNAL REGULATORY AND PUBLIC POLICY ATMOSPHERE

In considering small OP operations, it is important to consider the weight of the legal, regulatory, and policy atmosphere, which guarantees the autonomy of small OPs. The idea would be for OPs to be recognized as full actors—not as instruments of predesigned policies implemented without their consultation. From the perspective of local, regional, and national development, OPs have a key role to play; and in some countries, such as Peru and Brazil, coffee-grower cooperatives work continually through their representative institutions to shape public policies affecting the coffee sector.

In Colombia, coffee cooperatives and associations participate in coffee policy decisions through Comités Departamentales de Cafeteros (Departmental Coffee Grower Committees), with delegates taking concerns to the Comité Nacional de Cafeteros and Comité Ejecutivo (National Coffee Grower Committees and Executive Board), an advisory and decision-making body for the coffee sector. The most prominent new leaders also represent their associations at the Congreso Nacional de Cafeteros (National Coffee Grower Congress). Similarly, at the local and regional levels, cooperatives and associations have connected through constructive dialogue with municipal and departmental leaders to lobby for political and economic support for their development projects. In this way, and in contrast to some interpretations of the cooperative principle of "political neutrality," we consider involvement in defining policy both viable and appropriate as long as such efforts focus on the promotion of organizational interests and not on advancing the positions of individual leaders (see also Ruiz, this volume; Burke and Piekielek 2011).

COFFEE PRODUCERS' ASSOCIATIONS IN THE DEPARTMENT OF HUILA

Any discussion of Colombian coffee-producer cooperatives invites special attention to the region of Huila, a department located in the country's southwest. There, coffee-producer associations have gathered continuing strength, some with over two decades of history and others more newly formed. We believe that one of the most influential variables in the consolidation of this phenomenon is a regional identity based on mutual trust and a sense of loyalty—key characteristics in all associational work. In small OPs in Huila, we found an awareness among associates that collective success in negotiating and maintaining a direct relationship with the specialty coffee market depended upon the responsible functioning of their farms.

The specialty coffee program of ACDI/VOCA in Colombia funded by USAID began at the turn of the twenty-first century in five different departments. The department of Huila did not initially stand out for its coffee production or for its cup quality. In a few short years, however, Huila became a pioneer in the installation and staffing of quality analysis, or cupping, laboratories. It also distinguished itself by training qualified personnel to evaluate and harvest coffee in the highest-producing areas and by being the first to separate coffees according to their origin and quality. This constituted a step toward an initial characterization of producer subregions, which led to product differentiation with significant benefits for producers as well as for buyers.

The initial impulse for this transformation came from the Cooperativa Departamental de Caficultores de Huila (Coffee Growers' Cooperative of Huila, or CADEFHUILA), which marketed coffee from 26 of the 32 coffee municipalities in the department and enjoyed support from the FNC. The cooperative also started a training program for small-coffee-grower associations through a workshop called "High-Quality Coffee Production with a View to Exporting." Through this workshop, producers from more than 60 organizations got training. At the same time, the workshop was selected, supported, and promoted by ACDI/VOCA through its specialty coffee program, which started at the end of 2002. The effort later expanded to other departments in southwest Colombia, turning this region into the new frontier of specialty coffee production, with growth in terms of volume and quality and with origins identified and currently recognized by national and international markets.

By 2004, international market demands regarding quality and volume from coffee-producing countries had increased. This required improvements in coffee production and postharvest processing at the farm level—a step that pushed producers to adopt and incorporate new practices in order to capture better product value by negotiating special premiums for high-quality coffees (certified, sustainable coffees remained in an incipient state with very low volumes). Thanks to program-trained producers' adoption of new practices, international buyers offered economic incentives or special premiums for the quality of coffee produced. In addition, producers gained support for established regional socioeconomic and commercial development projects, which translated into immediate improvements in quality of life for their families and, as a result, their communities. In this particular case of cultural change and adoption of new practices, the economic stimulus was decisive; the parallel incentives of a higher selling price and the improvement of social and community life facilitated a process of "un-learning" in order to learn to negotiate a new commercial situation.

Change was noted in the participation of more family members in the coffee-producing process. Wives of coffee producers, who traditionally worked in the home producing some subsistence crops and generally taking care of domestic affairs, became empowered in administrative roles to manage record keeping, also gaining knowledge of agricultural best practices specific to coffee quality and directly participating in decision making regarding crop production and management. Young men and women within the farming household got involved in sensory quality control by participating in cupping workshops and later took charge of this process themselves in laboratories acquired by the OPs; they also participated in coffee quality control at the farm level.

Below, we present the case of one grassroots producer association that exemplifies some of the transformational processes described here. Its creation, evolution, and development during the last 10 years is discussed according to our perspective, which is a product of extensive collaboration and dialogue with association leaders and members.

THE OCCICAFÉ CASE

The Asociación de Productores de Café de Alta Calidad del Suroccidente de Huila (the High-Quality Coffee Growers' Association of Southwestern Huila, or OCCICAFÉ) is located in the La Plata municipality in the southwest of the department of Huila. Its 300 members live in 5 municipalities in which coffee accounts for 90 percent of the local economy. More than 10,500 family members have participated in OCCICAFÉ, working about 11,000 total hectares of coffee. Annual production of this region was around 20,625,000 kilograms of dried parchment coffee as a whole.[‡]

OCCICAFÉ was formed in Huila in 2002 as one of 14 associative groups that grew out of a state-sponsored training project for coffee growers who belonged to legally registered associations. The purpose of this initiative, which received technical assistance from the La Plata section of the Municipal Committee of Coffee Growers, was to offer training in issues related to sustainable coffee production and to evaluate the infrastructure needs of participating producers. As a contribution to the growth of high-quality coffee production, organizations received machinery and processing infrastructure at the farm level. For their part, the new OPs were required to comply with legal and tax obligations; this requirement arose as part of the national government's public policies and was one of the key points identified by the World Bank as an internal challenge to emerging organizations joining value chains.

A few months after the start of the project, organization leaders received official notification of its suspension, citing the government's insufficient budget. Although the news constituted a moment of frustration for OP representatives, faith in the initial proposal remained unaffected. The presidents of the 14 organizations met and analyzed the situation, together with the head of the La Plata section of the Municipal Committee of Coffee Growers, and reached the conclusion that their reasons for working together went beyond the purview of this project. They therefore set their sights on future production and marketing projects that would directly benefit their associations. This led in turn to the need for developing OP managerial capacities, another of the challenges identified by the World Bank.

Thus, OCCICAFÉ was founded as a formal and legal apex association in 2004, with the membership of the 14 grassroots associations, representing 264 coffee

[‡]This is the equivalent of 229,166 70-kilogram bags of Excelso coffee (green or golden bean), calculated with an average yield factor of 90 kilograms per bag, which is defined as the number of kilograms of dried parchment coffee needed to make a 70-kilogram bag of green bean coffee for exporting. This is the measure used in Colombia.

growers, legally constituted as voting members. This new organization was started under the leadership of one of the founding associations' presidents, who would go on to demonstrate clear advancement in management, technical aspects of production, input acquisition and distribution, marketing, and negotiations with other actors in the value chain.

From that moment, OCCICAFÉ opened its doors to serve its members, families, and communities, offering as its principal service the buying and marketing of coffee produced by associates and nonassociates who live locally. The organization grew rapidly, and by 2005, its membership had nearly tripled, including 601 associates and reaching 350,000 kilograms of commercialized dried parchment coffee.

Such rapid growth created problems in two domains. On the one hand, there was no clear notion of cooperativism, and a great many members failed to regularly attend meetings or uphold sustainability quotas, putting the organization's stability at risk. As a result, the president, with the board of directors and representatives of the founding associations, decided to rigorously apply internal statutes and regulations to disqualify and sanction those who were not meeting their legal obligations. They did not have to wait long for positive results. By 2006, the number of associates decreased to 381, but the marketing volume doubled, reaching 636,634 kilograms of dried parchment coffee. During the following two years, OCCICAFÉ marketed 1,660,614 kilograms of dried parchment coffee, making ever more patent its stability and maturity.

On the other hand, the rapid initial growth and the commercial activity of this OP prompted an immediate favorable response at the various coffee-buying points across the area, while marketing intermediaries rejected the unprecedented competition from producers. On several occasions, OCCICAFÉ had to invite local, departmental, and national coffee-sector representatives to show—by way of statistics as well as tangible results—that the commercial exercise in which they were participating would benefit everyone.

By centralizing storage in warehouses, the OP freed up private warehouse space and reduced labor costs in administration and in the handling of coffee during the intake process. Similarly, the coffee's transit through the central OCCICAFÉ site allowed for its evaluation and classification such that, when it was sold, it already carried a quality guarantee for the buyer—an exercise that improved odds of organizational sustainability.

From its start, OCCICAFÉ grew under the continued leadership of its president: a 40-year-old coffee producer with secondary school education, originally from the area, who proved himself a conscientious leader and worker on behalf of those he represented. Leaders like him illustrate three key factors for success: (1) fulfillment of policies, norms, and regulations that reflect the work of a credible, confidence-building board of directors or administrative council dedicated to organizational principles and teamwork with members; (2) study and knowledge of the market acquired through self-motivation and the process of understanding the organization's role within the value chain; and (3) the sense of belonging among members and members' families, in line with the organization's social, technical, economic, and commercial guidelines.

Currently, OCCICAFÉ member services include the following: (a) purchases of coffee at the market base price, fixed daily by the FNC and negotiated according to New York stock market fluctuations and trends through time; (b) future supplies and sales; (c) mechanical drying and monitoring of coffee through its various postharvest processes; (d) cup quality analysis tests and determination of the coffee's taste profile according to its subregion; (e) participation in quality contests; (f) specialty coffee and surcharge negotiations for identifying origin, for which the coffee is labeled *café de origen*; (g) loans for drying coffee; and (h) loans for sustaining and renovating coffee fields.

The participation of OCCICAFÉ in marketing the various qualities of coffee produced by its associates required developing new evaluation skills to satisfy the demands of its local, national, and international clients. To this end, the organization channeled its efforts toward obtaining and sustaining the Fair Trade Certified seal as well as the following certifications for individual members: Rainforest Alliance Certified, 4C (Common Code for Community Coffee), UTZ Certified, and the Starbucks C.A.F.E. (Coffee and Farmer Equity) Practices verification. Today, the OP has managed to occupy a space in the marketing chain that once was limited to private or intermediary marketers.

In terms of negotiation services around high-quality specialty coffee, OCCICAFÉ has achieved marketing with considerable price-for-quality differentials for large volumes produced by members, as well as for marketing for microlots of exceptional-quality coffee from specific associate farms adopting and maintaining agricultural best practices. These microlots were identified through cupping tests in the OCCICAFÉ quality laboratory and have garnered considerably higher-than-market prices thanks to the special interest of some international buyers. This constitutes a very specific and limited niche in the specialty coffee value chain.[§]

[§]The high quality of a coffee translates into a good yield factor not higher than 92.5 kilograms parchment per 70-kilogram bag of green coffee, which is reflected in the

During its existence, OCCICAFÉ has carried out various activities in the area of social development in conjunction with institutions such as the Comité de Cafeteros del Departamento de Huila and the coffee grower cooperatives supported by the FNC, as well as departmental governing entities, coffee exporters, and institutions like the Agencia para la Acción Social y la Cooperación Internacional and the ACDI/VOCA specialty coffee program, with resources from USAID. With support from ACDI/VOCA, OCCICAFÉ carried out additional activities that benefited families in relation to their farms, generating economic stability by increasing income, embracing women as farm managers, improving sustainability, and promoting best practices in production, respect for traditional crop management, food security, and environmental conservation. Similarly, sociocultural projects in rural communities led to resistance to internal conflicts and illegal crop growing (see Dávila, this volume).

Finally, it must be noted that in 2010, the department of Huila was recognized as the second-largest producer of coffee in the country. With 105,000 hectares of coffee cultivation, it garnered a bigger role in determining the FNC budget. Without a doubt, coffee growers in producer organizations in Colombia have absorbed the lessons of self-management and organization, as evidenced in Huila, where the seeds of future leadership have been sown.

FINAL CONSIDERATIONS

Sowing these seeds means, above all, recognizing the worth that producers and organizational leaders have in the country's coffee value chain. They are people who, despite a history of marginalization, have clearly proven capable of working collectively to make decisions that benefit their families and communities, empowering all those who have become proficient at different stages of the value chain, including women and young adults. Beyond the national and international arenas in which these associations are involved, their principal activity is selling what they produce, a goal that shapes the well-being and quality of life of their families, along with the stability and progress of the entire community.

table of minimum defects with the analysis of green coffee, a clean cup, and scoring on the SCAA quality scale corresponding to specialty coffees higher than 85 points, whose standards are accepted internationally and provide a common reference point in this market.

In Colombia and other Latin American countries, such undertakings represent the historic continuity of associations as instruments of local and regional development. None of what we have discussed would be possible through individual work alone, which is why the grassroots committees, producer committees, associate groups, associations, cooperatives, federations, and other legal denominations for rural OPs have an important reason for existing: community-based development.

In a similar fashion, while there are coffee growers in search of opportunities, so are there international cooperation and development projects, such as the one implemented by ACDI/VOCA, that aim to support them. It is important to make such project results known since fresh ideas and proposals from communities will advance the ongoing and necessary work of organizing rural producers in developing countries.

Coffee growers have learned their trade through inherited knowledge and practices, and their thinking remains creative, sensitive, and agile. Such was the case with those in the department of Huila, who were eager for opportunities and knowledge, and where members' capacities have known no bounds when presented with proposals that fit their vision. This vision fostered clear opportunities for creativity in order to achieve results supporting broad development and prompting the recognition of civil society and the community at large. A shift away from illegal crop production into the specialty coffee business through collective action and a value chain approach has liberated families from a violent context and has been conducive to the improvement of family dynamics, the empowerment of women, and the creation of leadership opportunities for young adults. The ability of these programs to remain strong requires strong organizations and solidarity at the local and global levels.

COLANTA

A Cooperative Unafraid of Change

LILLIANA ANDREA RUIZ MARÍN

ALTHOUGH MILK HAS COME TO FORM A FUNDAMENTAL PART of the Colombian diet, and milk production an important part of the agricultural economy, such was not always the case. In this chapter, I relate the story of how one humble and economically unstable cooperative managed to transform itself into a significant economic force and change the national milk market. The COLANTA cooperative, using a simple but well-executed economic strategy, provides a model for cooperative-led agribusiness development. Its history—and particularly the difficulties it has experienced because of Colombian agrarian policies—also shows the importance of developing a favorable policy context that supports rural producers and cooperatives. I draw on COLANTA archives, interviews with COLANTA members and neighboring nonmembers, and my experience working with other solidarity economy organizations to identify how this cooperative built a national market and addressed the challenges of trade liberalization.

INTRODUCTION

In 1962, a group of 64 small farmers in the Don Matías municipality (department of Antioquia) united to create the Colechera cooperative, with support from the departmental Secretariat of Agriculture. Their objective was to reverse their historical disadvantages in the dairy market, gain collective power by establishing economies of scale, and build the capacity to negotiate collectively rather than as individual producers. At that time, as indicated in the COLANTA historical archives, small milk producers in northern Antioquia faced numerous difficulties. Gold mining had stripped the land of its vegetation and fertility, and agriculture barely covered the subsistence needs of the families who depended on milk sales. Access to the fresh milk market in Medellín, the departmental capital, was nearly impossible since the mayor's office had prohibited sales of unpasteurized milk, and a local oligopoly controlled 95 percent of the dairy market. For consumers, milk was sold only every other day, and each family faced a maximum limit of two liters. Moreover, purchasing milk required long waits at neighborhood stores (COLANTA 2012b).

The Colechera cooperative became an object of ridicule in the department, derided for its seemingly utopian ideals and the fact that it declared legal bankruptcy three times in its first ten years. After the third bankruptcy, the government ordered the liquidation of the cooperative's debts, and many assumed that the story of Colechera was over. However, one year later the cooperative began a new cycle under the name COLANTA. Four decades later, COLANTA has taken its place as one of the most important dairy cooperatives in the country, contributing significantly to the favorable inclusion of smallholders in the national economy.

COOPERATIVES AND THE DAIRY SECTOR IN COLOMBIA

To assess COLANTA's impact on its members, it is important to understand the smallholder dairy sector in Colombia, which comprises individual producers with few resources and little political or economic influence. Land tenure patterns in the livestock sector are made up of smallholdings (*minifundios*), and families are heavily reliant on the milk they produce. According to the Colombian Livestock Federation, 82 percent of livestock production units (approximately 253,930 producers) have fewer than 50 animals (Lafaurie Rivera 2005). Dairy herds are even smaller and more fragmented, with an average of fewer than 10 animals. It is common to find families that milk 3 cows daily to obtain 12 liters of milk, and only 13 percent of producers in the country milk 15 cows on a daily basis, producing an average of around 50 liters and a monthly income of

900,000 Colombian pesos* (the equivalent of one and a half times the legal minimum wage in Colombia). These figures clearly illustrate the limited ability of dairy earnings to garner a fair wage for producers. Additionally, these small producers tend to be geographically dispersed in areas lacking sound public infrastructure. This leaves them with limited self-organization and little access to processing plants and markets, and thus economically vulnerable.

The dairy commodity chain is highly concentrated and does not favor small producers. Currently, there are approximately 140 processing plants in the country that supply fresh milk and dairy products to domestic consumers and international markets. Of the 499 milk collection and storage facilities in the country, 18 percent control more than 95 percent of total volume, and just 10 of these facilities account for 65 percent of milk storage (Castillo et al. 2005). This dairy processing structure leaves smallholders with minimal negotiating capacity vis-à-vis processors that deal with daily volumes in excess of 100,000 liters. In this context, a few cooperatives have played a key role in generating economies of scale, gaining power within the value chain, and stabilizing the household economies of small and medium-size milk producers. COLANTA has been particularly successful in this role.

While COLANTA members are similar to other Colombian dairy producers in the sense that the great majority (70 percent) are smallholders who have adapted their activities to the demands of the market, they receive a series of benefits that provide them significant and relatively unique improvements in their quality of life. The cooperative, with the overwhelming support of the membership, plays a significant role in ensuring quality standards from even the smallest and most geographically distant producers, guaranteeing that those small producers can engage in a profitable economic activity without having to abandon the countryside to migrate to the city. The support that a cooperative like COLANTA gives to the dairy sector is also critical at the macro level since the dairy sector has become important nationally, accounting for 4 percent of total GDP and 18 percent of food costs (Quintero Ospina et al. 2009).

---

*At the time of this writing, one U.S. dollar was equivalent to approximately 2,000 Colombian pesos.

COLANTA'S ECONOMIC SUCCESS

Since its rocky start, COLANTA grew to become a leader in the dairy sector, accounting for about 10 percent of Colombian milk production and managing a substantial sales volume at the national and international level (FEDEGAN 2005). The cooperative has a membership of more than 10,000 producers located in almost every region of the country, as well as 4,800 employees, who are also members (COLANTA 2012b). COLANTA boasts a line of more than 130 dairy and meat products and is the top cheese producer in Colombia, producing 36 percent of the national cheese supply and 48 percent of fresh cheeses (COLANTA 2010). In 2008, the cooperative registered $650 million in sales, making it the second-largest food company in the country (IAlimentos 2009). The economic success of the cooperative is such that, in many places, "going to COLANTA" (*ir a la COLANTA*) is synonymous with "going to the store." This success was achieved through an economic strategy focused on the following four principles.

ATTENTION TO PRODUCT QUALITY AND PRODUCERS' NEEDS

A central focus of the COLANTA cooperative has been increasing the quality of milk produced by its members and, consequently, the well-being of members themselves. To guarantee quality, the cooperative promotes the genetic improvement of Holstein breeds and gives free technical assistance to its membership through a technical department made up of veterinary doctors, animal technicians, agronomists, insemination specialists, and agricultural specialists. The cooperative incentivizes quality improvement by giving bonuses based on the protein content of milk, and simultaneously generates trust by guaranteeing the purchase of all of the milk that members produce. Finally, as the cooperative strengthens milk production, it also supports the economic diversification of member households with free business trainings for member families.

VERTICAL INTEGRATION

The vertical integration of the cooperative has allowed it to extend benefits to its members, reduce operational costs, and generate new sources of revenue. For example, the cooperative established a chain of fifty AGROCOLANTA warehouses, located strategically throughout Colombia to provide low-cost agricultural inputs and contribute to the broader solidarity economy (Salazar-Díaz 2009). Similarly, COLANTA makes and sells its own mineralized salts, fertilizers, and balanced feed rations. Through these programs, the cooperative has turned member needs into commercial opportunities and, at the same time, established itself as a recognized and trusted brand at the national level. At the other end of the value chain, the cooperative has established more than 100 of its own stores in cities across Colombia in order to reach consumers directly, without intermediaries. To extend its reach and establish another source of revenue, the cooperative has recently established franchises not only for product distribution but also for production itself. This strategy of concessions and franchises has served it well: in 2009, it saw a six-fold increase in earnings.

DIVERSIFICATION THROUGH INDUSTRIALIZATION

Innovation is the central pillar of COLANTA's continued business development. From its creation, the cooperative has invested heavily in the diversification of its dairy products and has used industrialization as a way of adding value to members' products. Currently, it has 38 dairy lines with 196 distinct products, 22 for meat, and 48 products in the line of soft drinks, juices, and bottled water. The cooperative intends to maintain its leadership in the dairy sector by developing innovative products, such as yogurt with Benecol, a bioactive ingredient used to reduce cholesterol (*Dinero* 2009).

MARKET EXPANSION

Making the most of members' increases in quality and production and of an efficient system for processing and transporting pasteurized milk, COLANTA managed to supply a previously unmet demand in Medellín, Colombia's third-largest city. Even more important for the cooperative, it established new national and international markets. In the last 40 years, COLANTA has played a fundamental role in the increase in Colombian consumption of milk and the acceptance of new dairy products. Figures provided by the cooperative indicate that Colombian per capita milk consumption grew from 50 liters in 1976 to 143 liters in 2008; the country transitioned from importing to exporting dairy.

In this period, the cooperative also reached a sales volume of 2.5 million liters per day (Colantapoli 2009).

In terms of new market incursion, it has opened doors to Russia, Chile, Mexico, Aruba, Costa Rica, Cuba, Ecuador, Morocco, Bangladesh, and Venezuela, exporting 1,200 tons of milk powder, yogurt, cheese, and cream each year, earning almost $12 million (*El Espectador* 2010). To access these markets, COLANTA has achieved or advanced toward a variety of strict international certifications, including ISO 1991,[†] the Russian qualifications for phytosanitary security (*El Colombiano* 2010), ISO 14001 for environmental management, OSHAS 18001 for occupational health and security, and HACCP for food safety. Similarly, COLANTA has been a pioneer in the certification project for Grade "A" PMO (Pasteurized Milk Ordinance) Milk Farms, which is indispensable for obtaining authorization for export to the United States (COLANTA 2012a).

These four economic strategies strengthen the institution and allow it to fulfill its social role with members and nonmembers. Faithful to its philosophy of guaranteeing total production purchase, COLANTA always has milk surpluses that it uses for donations and programs that distribute milk to schools jointly with the administration of the municipality of Medellín. In the last five years, COLANTA has invested $8 million in support of formal elementary, middle, and high school education; infrastructure such as the building of classrooms and sports complexes; and cooperative and technical training. Moreover, in the various municipalities where the cooperative is present, it has donated more than $4 million to support existing programs and activities, among them literacy programs for members and nonmembers.

Despite the cooperative's general pattern of growth, the balance sheet has occasionally suffered in the past several years. COLANTA lost $3.5 million in 2002, and it registered losses of $9.1 million in 2005. That same year, the company lost nearly 8 percent of its net worth, dropping from $71.4 billion in 2004 to $65.8 billion in 2005 (*El Tiempo* 2006). Opponents of the cooperative's administration argued that these declines were due to the president dedicating excessive time to political activities as a Colombian congressman. With the

[†]ISO is the International Organization for Standards, which establishes global standards, including food safety, environmental management, and occupational health and safety.

generalized crisis of 2007–8, a small fall in internal demand, and export restrictions to Venezuela, the cooperative had to assume the additional costs of storing the milk it buys from members (Portal Lechero 2010).

In 2010, the mayor of the municipality of San Pedro de los Milagros and nine municipal council members attempted to rescind COLANTA's commerce and industry tax exemption, from which the cooperative had benefitted for 23 years—the full time that the cooperative had been present in the municipality. According to the mayor, in just the preceding seven years, the municipal treasury had forfeited nearly 6 million pesos in revenue through these tax exemptions (Echeverry Marín 2010). The council members argued that the billions of pesos that COLANTA was exempt from paying each year could be invested in the community.

Despite COLANTA'S recent economic setbacks, the national dairy policy promoted by the National Council of Social and Economic Policy (Consejo Nacional de Política Económica y Social, or CONPES) has adopted COLANTA's path to success. In June 2010, CONPES published a document intended to improve the competitiveness of the Colombian dairy sector by lowering production costs, raising productivity, diversifying internal and external markets, and making the most of opportunities and comparative advantages within the sector. Among its objectives are the promotion of agro-industry, associativism, horizontal and vertical integration, processing and sale of products with value added, improvements in quality, and the strengthening of institutional management (CONPES 2010). Put simply, the new recommendations suggest that everyone should follow COLANTA's model.

THE CHALLENGE OF UNFAVORABLE POLICIES

My interviews with cooperative members, ex-members, and people who belong to other cooperatives point to a sharp critique of the Ministry of Agriculture for not having developed a national dairy policy that would involve diverse stakeholders. Several of those interviewed explained that the government has legislated to exercise control, but that in the majority of cases, this has favored the country's large dairy processors with little consideration for the milk producer. As I explain below, the COLANTA cooperative, an entity with a double identity as both large seller and representative of small producers, has suffered as a result of different local and national policies during its history.

The first policy challenges involved trade liberalization and international competition, often under terms that were disadvantageous to Colombian smallholders. In 1979, the national government authorized the import of 43,000 tons of powdered milk at below-market prices—an enormous increase given that the historic average between 1967 and 2002 was only 7,518 tons per year. This situation reduced the cooperative's buying capacity and forced it to give away glasses of milk since there was no way of getting powdered milk into the market (COLANTA 2011). The lack of competitiveness of the Colombian dairy sector is linked to low productivity, high production costs, and competition with subsidized international markets (Lafaurie Rivera 2009). The cooperative, with its guarantee of buying all the milk its members produce, protects smallholders from the harmful effects of this type of policy, but in the process, the institution itself assumes the costs and risks that go along with these policies.

A more recent crisis has to do with the import of whey, which went from no imports to an average yearly importation between 1981 and 2002 of 1,184 tons. Starting in 2003, the government allowed increases so that by 2010 imports had grown five-fold, reaching an average of 7,423 tons (*Portafolio* 2011). Whey powder is used as a sugar substitute in the preparation of baked goods and chocolate. Leaders of the dairy sector believe that these imports are unnecessary because Colombia has enough supply to satisfy the market, yet these policies force them to compete with foreign dairy farms.

A second key policy issue has been national price controls for milk. COLANTA members are always concerned about prices, particularly because the Ministry of Agriculture has repeatedly changed the protocol for paying for a liter of milk. For many years, COLANTA had established bonus payments for quality linked to protein content. In 2007, the Ministry of Agriculture established another way of defining price, according to which the industry was to pay producers a composite base price (which was adjusted every six months) plus voluntary bonuses for volume, refrigeration, and seasonality, and mandatory bonuses for hygienic and sanitary quality. According to COLANTA, this new formula minimized the importance of protein and fat—that is, the useful parts of the milk—an error that was demonstrated with the decrease in average protein content, which fell from 3.2 to 3 percent between 2003 and 2009. Moreover, the new government pricing ignored the facts that the international price of powdered milk varies and that there are periods of rain and drought that affect production, supply, and pricing. Spurred by disagreement among government analysts, the various livestock unions, and the cooperative sector, the government has passed two other price-control resolutions in recent years, increasing farmer concerns and disrupting COLANTA's financial incentive system that rewarded quality and technical improvements.

Finally, the implicit acceptance of informal markets also harms COLANTA and its members. According to the Colombian Livestock Federation, the informal milk market enjoys certain advantages in comparison with the formal market. It is estimated that the informal market sells 40 percent of the national fresh milk supply. The informal sale of fresh milk is a business with a high profit margin, characterized by low operation costs and high cost to consumers (close to that of pasteurized milk), and it is has benefited from reductions in farm-gate prices since consumer prices have never gone down (Lafaurie Rivera 2005). In September 2008, the Ministry of Agriculture made the problem of informal markets worse by reestablishing the sale of raw milk.

CONCLUSION: CHALLENGES FOR A FUTURE OF FREE TRADE

The forty years of work that COLANTA has invested in its strategies for continual improvement in the areas of productivity, quality, diversification, agroindustry, and sales have changed the Colombian dairy market. These strategies have enabled the cooperative to transform—at least for its members—a sector that was once characterized principally by highly dispersed and marginalized producers and highly concentrated and powerful dairy processors. This new market power, in turn, has allowed the cooperative to make the most of the commercial opportunities brought by trade liberalization and economic globalization. Among other indicators of its success, COLANTA stands out as the largest dairy cooperative in the country, the result of its constant work to consolidate the microeconomies of thousands of small-producer families.

The success of the cooperative is evident despite a number of setbacks related to state policies that have left small dairy farmers walking a tightrope between stability and ruin. Moreover, with its large market share and capacity for management, the cooperative has recently won the ear of decision makers who directly shape the future of the sector. Now the new free trade agreements that Colombia has or will put into place will give COLANTA new opportunities while simultaneously exposing the cooperative to competition with the largest dairy companies in the world.

Many studies predict the negative impact of free trade on Colombian smallholders, and, although dairy producers probably will not feel the most serious impact, they will not be exempt. According to the Colombian Action Network against Free Trade:

> It is calculated that, with all the [free trade agreements] that Colombia has adopted, some 15,000 tons of powdered milk will be allowed to enter the country every year, when during 2009, the national dairy industry maintained inventories of approximately 20,000 tons of unpurchased powdered milk. In addition, quotas of whey from Switzerland, Norway, Iceland, the European Union, and the United States in the amount of more than 10,000 tons will enter, while the demand of the Colombian market is at 2,000 tons and entry quotas in the various FTAs were authorized at 5,000 tons. (RECALCA 2010)

Beyond the supply imbalances that these agreements will engender, it cannot be ignored that the European Union and the United States provide significant subsidies to their producers, which makes any type of international competition difficult.

Milk producers and unions strongly opposed these free trade agreements. In response, the national government committed $282 million to protect and strengthen the competitiveness of the dairy sector over the 17 years of transition to free trade with Europe. CONPES argues that this level of investment is insufficient, amounting to only $16 million per year. The Colombian Livestock Federation has projected a minimal level of support at $80 million per year in order to adapt to the coming economic environment (González 2010).

Still, the COLANTA cooperative insists that it is prepared to meet these challenges and that it can successfully enter foreign markets, principally in the United States. Perhaps history supports its confidence. Castillo et al. (2005) emphasized COLANTA's success in making the most of the MERCOSUR regional market—an achievement due to its historic dedication to quality, production, and diversification. While the free trade agreements present many difficulties at the national level, the primary difficulty for entering new markets will be meeting phytosanitary standards. Since COLANTA already has or is about to receive the certifications required by several importing countries, it is well positioned to expand its participation in international markets. According to Santiesteban Rojas, "The effects of the FTA [with the United States] threaten above all the smallholder or informal agriculture economy whose products are destined for internal sales" (2008, 61). Garay Salamanca et al. (2009) also offer some hope. Although the proposed free trade agreement clearly favors the United States over the first 15 years, "in the long term, the terms agreed for [the dairy] product chain may turn out to be favorable for Colombia, given the relative size of the markets, and the possibility that the Colombian dairy industry may compete successfully in the United States" (36). The leaders of COLANTA have demonstrated resilience in facing these challenges in the past, and they are confident that their strong cooperative foundation and commitment will be strong enough to meet the challenges now looming on the horizon. What is certain is that, of all the small dairy producers in the country, COLANTA's members have the best possibility of surviving and riding this new wave of change.

RURAL SAVINGS AND LOAN COOPERATIVES

Five Cases in the South of Santander

SONIA CAROLINA LÓPEZ CERÓN

THE FIRST RURAL SAVINGS AND LOAN COOPERATIVES (Cooperativas Rurales de Ahorro y Crédito, or CRAC) in the southern region of the department of Santander, located in the northeast mountain range of the Colombian Andes, emerged in the 1950s and '60s. As in other regions of Latin America, these first cooperatives were promoted by the Catholic Church, which developed a strong policy of inspiring solidarity and a culture of savings among the peasantry. Their work was boosted by agricultural trade unions that simultaneously drove the formation of agro-industrial cooperatives dedicated to production (e.g., dairy cooperatives) or commercialization (e.g., coffee and sugarcane).

In the case of southern Santander, support from the Catholic Church was offered through the Secretariat of Social Ministry (SEPAS) and later strengthened with the creation of the Central Cooperative for Social Promotion (COOPCENTRAL). This program, referred to as SEPAS-COOPCENTRAL (Coque Martínez 1996, 64), aimed to promote development by establishing a cooperative in every municipality in the region. In the words of the Catholic priest Ramón González,* the director of SEPAS and the prime mover of the

*Father González had carried out studies on cooperativism and rural development in Belgium, Italy, and Canada at the beginning of the 1960s. He put these studies into practice through SEPAS-COOPCENTRAL (Coque Martínez 1996).

SEPAS-COOPCENTRAL model, the creation of cooperatives "gave the organization a legal point of departure and, at the same time, stability" (Dávila and Silva 1996, 89).

Building on studies carried out by the Solidarity Studies Unit of the School of Rural and Environmental Studies at the Pontificia Universidad Javeriana, I present the most important development outcomes from five such cooperatives supporting the smallholder economy of southern Santander. In so doing, I synthesize research on these cooperatives conducted since the 1980s. I analyze these cases from the point of view of their contributions to local development, focusing on four key elements: local cooperation, social learning processes, social capital and leadership, and the creation of independent capital.

THE SOUTHERN REGION OF SANTANDER

The southern region of Santander comprises three of the six provinces[†] that make up the department of Santander: Guanentá, Comunera, and Vélez. Together, these cover approximately 12,000 square kilometers with close to 500,000 inhabitants living in 53 municipalities, of which only 4 have a population greater than 50,000 residents. A diverse rural economy predominates in these provinces, with small and medium landowners as well as landless peasants (sharecroppers and day laborers) who work in production, services, handicrafts, commerce, and transportation. These provinces are poorer than the overall departmental average. For Santander as a whole, 22 percent of the population is described as having unmet needs; however, this percentage rises to 40 percent in Vélez, 33 percent in Comunera, and 29 percent in Guanentá (Plan de Desarrollo 2008, 61). Public infrastructure is also limited in southern Santander. While across the department, 21 percent of roads are paved, the percentage falls to 12 percent for Vélez and 16 percent for Comunera (Ibid., 148).

The provinces served by SEPAS-COOPCENTRAL are located close to the Middle Magdalena River Valley and the city of Barrancabermeja, home to the largest oil refinery in Colombia. This has been one of the most troubled areas of the country, characterized by violent confrontations among paramilitary groups,

[†]According to Colombia's 1991 Political Constitution, provinces are territorial entities made up of municipalities, and municipalities can be subdivided into urban neighborhoods (*comunas*) and rural districts (*corregimientos*).

guerrillas, and the national army. These groups are the primary representatives of the armed conflict in Colombia, which has lasted for more than forty years. The causes of conflict include struggles for land control, social inequality, and drug trafficking. It is important to emphasize that, despite the very difficult situation facing communities in southern Santander, the three provinces analyzed in this study have managed to remain at the margins of the conflict and achieve some level of peace and stability. As Coque Martínez explains in a study from 2005, cooperative leaders attribute the relative peace of the area to the development impacts generated by the cooperatives.

The development process promoted by SEPAS and COOPCENTRAL during the 1960s focused on treating provinces as natural development "territories," identifying and training natural leaders, organizing the population through a network of grassroots cooperatives, and promoting in people's minds the notion of an achievable utopia (Bucheli 2006). The network of first-tier and second-tier cooperatives is currently made up of 17 savings and loan cooperatives (CACs, or Cooperativas de Ahorro y Crédito), of which 13 are located in municipalities with fewer than 12,000 inhabitants and are classified as CRACs (Cooperativas Rurales de Ahorro y Crédito) since they are in rural areas. As indicated in table 16.1, the oldest of these rural cooperatives was created in 1952 and the youngest one in 1984; however, the majority arose in the 1960s and '70s under SEPAS and the guidance of Father Ramón González. As seen in table 16.1, the membership has increased significantly in all the cooperatives except for Guadalupe. This growth reflects the regional development impacts of these cooperatives (Dávila and Forero 1997, 29).

Santander has a proportionally large number of the cooperative banking institutions nationally. Of the country's 192 CACs and 7 other financial cooperatives, 13 percent are located in Santander, and only Bogotá and Antioquia account for more cooperative banks.[‡] In terms of the rural savings and credit cooperatives (CRACs), the majority are located in the departments of Antioquia and Santander (10 CRACs each). These cooperatives serve the needs of rural producers, although some of the urban-based CACs also offer services to rural members.

[‡]Whereas CACs are permitted to manage resources only from their associates, whether through contributions or savings, the other financial cooperatives can manage resources from third parties as well.

TABLE 16.1. Founding Years of CRACs and Evolution of Number of Members, South of Santander

COOPERATIVE, MUNICIPALITY	YEAR	NUMBER OF MEMBERS		
		1985	2005	2009
Barichara Multiservice Cooperative, Barichara	1961	1,556	3,800	4,459
Agricultural Cooperative Encino, Encino	1987	N/A	224	226
Multiactive Agricultural Guadalupe, Guadalupe	1962	52	2,071	1,843
Credit and Savings Cooperative La Belleza, La Belleza	1961	290	848	1,238
Integral Agricultural Cooperative La Paz, La Paz	1984	29	1,001	1,340
Multiservice Cooperative Mogotes, Mogotes	1952	805	4,367	7,200
Integral Peasant Cooperative Paramo, Paramo	1972	77	657	841
Puente Nacional Credit and Savings Cooperative, Puente Nacional	1969	1,062	1,555	3,381
Colombian Financial Cooperative for Solidarity Development, Socorro	1962	2,136	24,131	62,761
Credit and Savings Cooperative La Granja, Sucre	1974	390	428	690
Credit and Savings Cooperative Valle de San Jose, Valle de San Jose	1967	550	4,416	5,372
Multiple Services Cooperative of the Vélez Province, Vélez	1968	1,494	3,440	6,009
Multiple Services Cooperative Villanueva, Villanueva	1969	1,919	8,602	8,896

SOURCE Original table, composed with data from CONFECOOP. Data for 1985 taken from Dávila and Forero 1987, 29.

TRUST, PARTICIPATION, AND SAVINGS

By comparing trends in internal financial operations, it is possible to assess such factors as organizational trust, participation, and member savings with CACs at the national and departmental levels, as well as among the thirteen CRACs in southern Santander. Changes in the total value of the sector loan portfolio (in 2008 dollars) are depicted in figures 16.1a, 16.1b, and 16.1c. The component parts of the portfolio value are the contributions of members as they join the cooperative and the deposits they make in the form of savings. Increases in the portfolio values indicate a growth in demand for credit, while the higher proportion of deposits to contributions reflects a tendency toward savings. Santander's CACs (fig. 16.1b) have behaved similarly to national CACs (fig. 16.1a) but portray a higher rate of growth and a lower deposit-to-contribution ratio. Still, the sum of contributions and deposits can cover the full loan portfolio, meaning that these cooperatives can finance their activities with low-cost resources, which translates into lower interest rates for members. The portfolio-deposit-social contribution behavior for the department is similar to that of the group of 13 CRACs in southern Santander (fig. 16.1c). In every case, the average loan portfolio growth is greater than the average growth of deposits and social contributions. Nevertheless, the difference between the growth in deposits

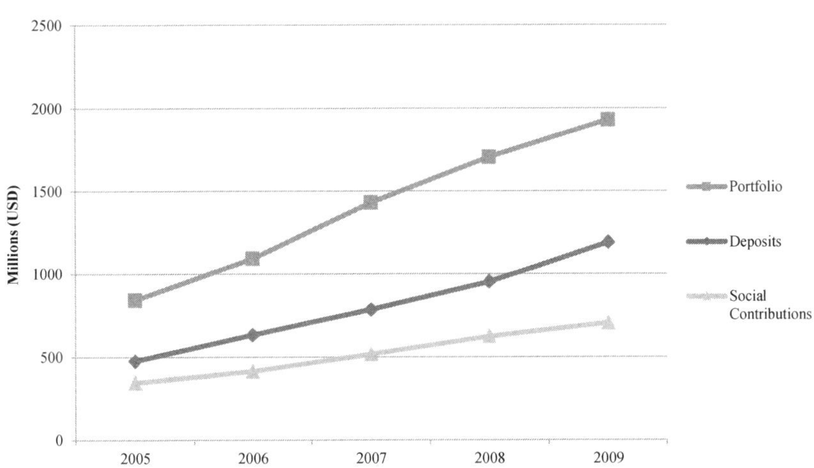

FIGURE 16.1A. Financial operations in CACs across Colombia

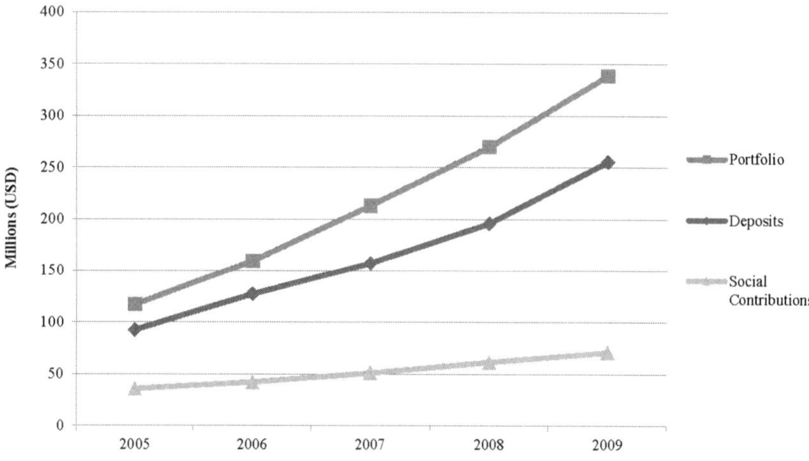

FIGURE 16.1B. Financial operations in CACs across Santander

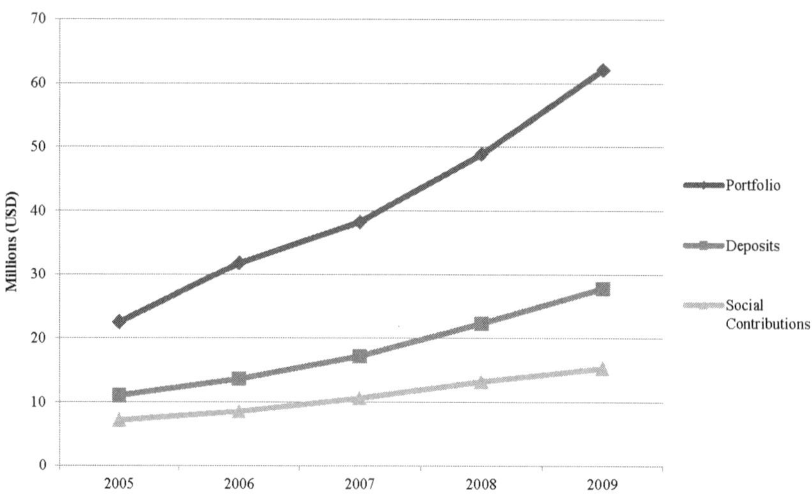

FIGURE 16.1C. Financial operations in 13 CRACs in southern Santander

and the portfolio is smaller for the Santander area and for the 13 CRACs. The strengthening in deposits is fundamental for a CAC because it reflects the level of financial savings in a population.

The growth of these financial operations demonstrates the strength and sustainability of these cooperatives. The trust placed in the organization by mem-

TABLE 16.2. Four Indicators of Financial Standing: Ratios of Portfolio to Assets, Deposits to Liabilities, Member Contributions to Patrimony, and Profit to Patrimony

	PORTFOLIO/ ASSET	DEPOSIT/ LIABILITY	CONTRIBUTIONS/ PATRIMONY	SURPLUS/ PATRIMONY
Total CACs Colombia	80.6%	79.0%	71.6%	6.9%
Total CACs Santander	78.2%	86.8%	62.0%	7.2%
Total 13 CRACs Santander	76.9%	92.7%	59.6%	8.7%

SOURCE Original table with data from CONFECOOP.

bers is reflected in the deposits as well as in the social contributions, with the latter being the fundamental driver of a collective-action organization. The loan-to-assets ratio presented in table 16.2 indicates that Colombian CACs have a high level of active loans, which is to be expected given that this is their primary activity. At the same time, deposits are the largest segment of total liabilities. This indicator is higher for the group of 13 CRACs, which is to say that they have a lower degree of indebtedness to third parties and may therefore be able to provide lower interest rates for their associates. Finally, the ratio of member contributions to assets is higher among the provinces compared to the national data. Additionally, the ratio of surplus to assets is stronger for the group of 13 CRACs as well as for the entire department. Accordingly, it is possible to conclude that the CAC and CRAC sectors are financially sound and present positive economic growth.

FIVE CASE STUDIES IN THREE PROVINCES

The five most-studied CRACs of the province have been selected as the case studies for this chapter: Cooperativa de Ahorro y Crédito de Valle de San José (COOPVALLE); Cooperativa Agropecuaria Multiactiva de Guadalupe (MULTICOOP); Cooperativa de Ahorro y Crédito de la Granja (COAGRANJA); Cooperativa Multiactiva de Servicios de Barichara (COOMULSEB); and the

Cooperativa de Ahorro y Crédito para el Desarrollo Solidario de Colombia (COOMULDESA). The essential indicators of comparison of CRAC performance and development success are local cooperation, social learning outcomes, social capital and leadership, and independent capital (Dávila 2008; Lobo 2002). With these qualities, CRACs can play a key role in processes of local development, being alternative organizations whose purpose is service and quality, whose philosophy facilitates reciprocity, autonomy, and loyalty based on trust, and whose operations empower stakeholders (Ramírez 2002).

THE IMPORTANCE OF LOCAL COOPERATION

Development projects are shaped by the historical and cultural contexts of each locality, and cooperatives, as collective organizations, have the potential to maximize the redistributive impacts generated by the projects in their zones of influence (Coque Martínez 1996, 66). CRACs especially contribute to local development solutions thanks to three unique features of their organizational model: their rootedness in the local social fabric, their building of resource reserves, and their promotion of mutual cooperation (Ibid., 67). Based on these three factors, CRACs generate the independent capital that consolidates their organizations and inserts them into networks that make possible the exchange of knowledge and resources that meet the needs of their membership in ways that could not be achieved individually.

In the case of SEPAS-COOPCENTRAL, this process began in the 1970s, offering managers of affiliated cooperatives training in the areas of auditing, consulting, and loan monitoring (Dávila 2002, 71). This initial training allowed the CRACs to develop a financial system capable of improving the smallholder economies of their members. As Dávila explains:

> One of the great strengths shown by CRACs in southern Santander . . . is the capacity to link this service provision with the recognition of [local] social and economic characteristics, offering services that might be called "tailor-made" according to the social and economic characteristics of their members . . . In this way, loan services are offered with timeliness, flexibility, and understanding—elements that traditional financial intermediaries are not always able to offer as they serve a vision that does not consider the local cultural context, preferring rather

to minimize operational costs or take advantage of economies of scale. (Dávila 2008, 316)

As López and Peña (2005, 4) have observed, financial services for the smallholder in Colombia are insufficient, given that a good portion of the municipalities with large rural populations have limited access. In fact, almost a third of municipalities in the country have no financial services of any kind, and in the remaining municipalities, the small-producer population has highly limited access to credit. In southern Santander, loan services are provided through local CRACs or those in neighboring municipalities where traditional financial intermediaries are unavailable. For example, COOMULDESA today has offices in 13 municipalities in the region. COAGRANJA is the only financial entity in the municipality of Sucre, where its services are appreciated by the community because they respond to their needs and are provided more efficiently than through the traditional financial institutions (Dávila and Forero 1987, 36).

In addition to providing savings and loan services, these cooperatives have developed other services tailored to the needs of associates and to the community in general. COOPVALLE, for example, offers a line of credit at low interest rates and technical assistance for member farm projects. In this way, COOPVALLE has been able to support the primary agricultural activities in the municipality (cattle raising, pig farming, and *panela* sugarcane cultivation). COOMULSEB has contributed to technical changes in bean cultivation through the acquisition and rental of machinery, including a bean thresher. Bean cultivation in the municipality of Barichara is one of the most important production activities and has replaced tobacco as the principal cash crop.

MULTICOOP is located in a municipality dependent on dairy cattle, and the cooperative provides artificial insemination services and has created the infrastructure for marketing milk in a way that guarantees high prices and price stability. COAGRANJA established a consumer section that offers wholesale and retail consumer products to their membership at attractive prices, thus expanding the benefits of collective action. Even when existing commercial establishments lowered their prices in response, the local population has preferred to buy from the cooperative store (Ariza and Lobo 2002, 320). For its part, COOMULDESA created the COOMULDESA Foundation in 1997 when regulations that govern the solidarity economy sector began to limit the actions of multiactivity CACs and to require that they follow rules similar to those for banks. Through this foundation, COOMULDESA provides marketing services for agricultural products such as corn, beans, and coffee (Bucheli 2002, 283).

CRACs, on the other hand, tend not to reach the most vulnerable segments of the population. Low-income households can find it difficult to raise the membership fees and deposit the savings amounts required to join a CRAC. These families are therefore served primarily by microfinance mechanisms that accept nonmonetary guarantees to provide access to credit for people without access to traditional financial institutions.

COOPERATIVE SOCIAL LEARNING OUTCOMES

The development model of the region has relied on cooperative organization to provide the mechanism for enhancing managerial capacities among the membership. This effort has involved the training of leaders who become cooperative managers and the promotion of a culture of participation within management, assembly meetings, and everyday decision making. As a result, in each cooperative it is possible to identify members who have taken an interest in learning and placed themselves at the disposal of the cooperative.

Some cooperatives, such as COOPVALLE and COOMULSEB, have had low turnover among their managerial staff, which has slowed participation and the growth of managerial skills among the wider membership base. COOMULSEB successfully addressed this problem in 1993 by changing leadership election processes in ways that increase participation and decrease the concentration of power (Dávila 2002, 230). In the cases of COOMULSEB, MULTICOOP, COOMULDESA, and COOPVALLE, managerial capacity-building was especially pronounced in the areas of decision making and evaluation, creating what Dávila has called "a permanent process of trial and error, a process that has generated efficacy and efficiency in management of the organization" (Dávila 2004, 70). MULTICOOP has also developed leadership capacities, manifested in the enthusiasm with which the manager "has taken to directing, orienting, and motivating the creation of new ideas, with the end goal of converting them into new services" (Ibid., 105). At COAGRANJA, the fact that associates have learned to "do math" stands out, as does the awareness of the importance of saving, which resulted in an increase in deposits (Ariza and Lobo 2002, 330).

Women are also increasing their roles in cooperative management. COOMULSEB has seen an increase in women's participation on the Administrative Council, from 14 percent of council members in 1985 to nearly 67 percent in 1999 (Silva and Dávila 2002, 232). It should be noted that a woman, consistently elected since 1975, currently occupies the post of manager. In the case of COOPVALLE, an increase in women's participation was fundamental for instituting an inclusionary strategy driven by collective concerns and the clear goal of achieving benefits for all associates (Dávila 2004, 80). This change is also reflected in the cooperative's leadership; since 1993, a woman has occupied the post of manager. Women's participation in the Administrative Council remained between 20 and 40 percent for the 1991–2001 period but has increased in the assembly of delegates, going from 27 to 46 percent over 10 years.

SOCIAL CAPITAL AND LEADERSHIP

Social capital is critical for building values, norms, and networks of trust, and cooperatives play a leading role in creating the networks that drive local development (Lobo 2002, 138). In all the cases under consideration, management has demonstrated its dedication to the values and principles of the cooperative doctrine. In this way, it has favored a culture of self-control, honesty, transparency, and ethics (Bucheli 2002, 278). COOMULSEB, MULTICOOP, and COOPVALLE are private organizations where "concern for the general well-being and for the community reigns" (Dávila 2002, 236). In COOPVALLE, there is an awareness of the importance of the role that the cooperative plays in the social and economic life of the community, and for that reason it seeks "to create favorable conditions for carrying out its production activities and social and economic duties under the best possible conditions" (Dávila 2004, 63).

COAGRANJA has contributed to local development by building a warehouse to reduce the price of farm inputs and increase the availability of financial services tailored to local needs. It also played an important role in securing peace in the town through its participation in a process of changing values and behaviors. In October 1997, the La Granja and Sabana Grande parishes declared themselves "peace territories" (Ariza and Lobo 2002, 336). Additionally, in association with the Catholic Church, they mobilized the community to obtain a local bus—an important service given the distances and poor quality of roads within the municipality (Ibid., 348).

BUILDING INDEPENDENT CAPITAL

In order to offer loan services, CRACs in southern Santander use the deposits and contributions of their members (López and Peña 2005, 42). By drawing capital from the cooperative's own resources—that is, from member fees, deposits, and profits—they are able to keep capital costs extremely low. For example, membership contributions have a cost to the cooperative approximately equivalent to the annual rate of inflation, and savings deposits are cheaper than bank loans (via the interest rates that banks charge). Cooperative capital stocks have costs. For example, in the third trimester of 2010, the cooperative, through a fixed rate of deposit, paid about 5.3 percent for the use of member capital (CONFECOOP 2010, 16), well below the cost of capital borrowed from the formal banking system.

Independent capital is accrued when cooperatives charge associates fair interest rates in accordance with the cooperative philosophy, although capital is often scarce compared with the credit needs of CRAC service beneficiaries. Dávila's (2008) study collected information from 2007 to show that 15 of the 17 CRACs had approximately $50 million available to provide credit and other services to the membership. Thirty percent of the capital came from the member contributions (fees), and 70 percent came from member savings deposits, of which almost $31 million had been invested (Dávila 2008, 310). With this internal capital stock, CRACs offered four lines of credit: agricultural, commercial, consumer, and microbusiness—following the priorities established by the directives imposed by the Superintendency of the Solidarity Economy (SES), which is in charge at the federal level of the supervision and control of solidarity entities (employee, cooperative, mutual, and other funds). It is important to emphasize that the SES rules set during the 1990s differ from the established credit practices promoted by CRACs, which set credit lines and interest rates to respond to membership needs. In all the cases presented here, the average interest rate charged by the cooperatives was under the maximum legal level of 27.4 percent set by SES (López and Peña 2005, 35).

The growth of internal capital stock has been the result of a sustained effort. It is true that COOMULSEB, COAGRANJA, COOMULDESA, and MULTICOOP have experienced periods of financial difficulty marked by insufficient capital to cover the demand for credit. Nonetheless, they have weathered these crises, and the key indicator of sustainability is the fact that these cooperatives have managed to strengthen the loyalty of their members, who supported their organizations en masse during the severe crisis of 1997–1999 in Colombia. During this period, 49 large, primarily urban financial cooperatives, along with three cooperative banks, were dissolved as a result of several factors, including corruption, poor management, and lack of discipline. The lack of intervention by government financial institutions also played a decisive role by not facilitating access to the necessary capital to those cooperatives that were well managed (Dávila and Bacheli 1998, 31–35). In the case of the southern Santander cooperatives, however, there was not a mass exit of members or a loss of faith in their organization.

CONCLUSION

CRACs have played an important development role in southern Santander by generating social learning processes, social capital and leadership, and independent capital. As these cooperatives have developed internally, they have been increasingly able to promote rural development tailored to the communities they serve, and they have strengthened community ties and community capacities. The five CRACs analyzed in this chapter have provided credit services at reasonable rates of interest, expanding coverage to meet the demand for credit. In addition, each has developed services that address the specific needs of each locality, contributing to farm systems, the training of leaders, and the growing participation of women in managerial posts. A description of the development contributions of each case study is provided in table 16.3.

It is important to emphasize that the successful experience of these five CRACs unfolded in a context of (and despite) public policies that tend to limit the reach of these cooperatives into their respective communities (Burke and Piekielek 2011; Gutiérrez 2014). On the one hand, the regulation of the solidarity sector at the end of the 1990s began to limit the ability of multiactive CACs to act, insofar as they are required to honor regulations similar to those structuring the banking industry. On the other hand, high levels of rural poverty related to land dispossession and displacement caused by violence do not allow the most needy people to have the minimum resources required to be able to join a cooperative and access its services.

It should also be remembered that these cooperatives have managed to create and contribute to the maintenance of a peace territory in the midst of

TABLE 16.3. Contribution to Local Development: A Four-Factor Analysis

	LOCAL COOPERATION	SOCIAL LEARNING	SOCIAL CAPITAL AND LEADERSHIP	INDEPENDENT CAPITAL
COOPVALLE	1. Development of services that respond to member needs (lines of credit for production projects, with technical assistance) 2. Management of cooperation projects (DEVALAR project)	1. Low turnover in managerial posts 2. High participation of women in managerial posts	1. Leadership shared among several people 2. Collective management 3. Difficulty working with the municipal mayor's office	Sufficient and based on deposits
MULTICOOP	1. Development of services that respond to the needs of associates 2. Creation of other services (purchasing milk from the community in order to sell it)	1. High turnover in managerial posts 2. High participation of women in managerial posts	Leadership centralized, but with a collective management approach	Sufficient and based on deposits

COOMULSEB	1. Development of services that respond to the needs of associates (machinery rental, bean seed thresher, contributing to the technical improvement of bean cultivation) 2. Continual presence in cultural and recreational activities	1. Moderate turnover in managerial posts 2. Growing participation of women in managerial posts	1. Leadership shared among several people 2. Promotion of healthy competition for offering fair prices for products needed by the community	Sufficient and based on deposits
COOMULDESA	1. Development of services that respond to the needs of associates 2. Creation of the COOMULDESA Foundation	Low participation among the base	Leadership centralized in management	Sufficient and based on deposits
COAGRANJA	1. Active in the peace process in the area 2. Management of cooperative projects 3. Creation of other services for the community (consumer department for regulating prices)	Moderate turnover in managerial posts	1. High participation in activities led by local organizations 2. Promotion of healthy competition for offering fair prices for products needed by the community	Sufficient and based on deposits

the conflict suffered in Colombia and particularly in the nearby region of the Middle Magdalena River Valley. This provides an important counterpoint to research showing that Colombian armed actors use grassroots development projects to facilitate violence, land grabbing, dispossession, and the construction of shadow states (Ballvé 2013). My conclusion is that by working collectively, training leaders, generating social capital and independent capital, and making employment and self-employment possible, CRACs strengthen their members and communities. Further study should help to deepen our understanding of how small-producer cooperatives empower rural communities to confront the dominant sources of vulnerability and marginalization.

CONCLUSION

COOPERATIVES AS CHANGE AGENTS IN RURAL LATIN AMERICA

Synthesizing Experiences Across Countries

MARCELA VÁSQUEZ-LEÓN AND BRIAN J. BURKE

LATIN AMERICA HAS SEEN SIGNIFICANT transformations in the last several decades, sparked in large part by nearly 70 years of international development, programs of land reform and state-supported rural development, the collapse of authoritarian regimes, the advance of increasingly inclusive democratization, the rise and fall and rise of left-leaning governments, and neoliberal policies supporting economic globalization. Rural Latin America surely looks different today than it did in the 1950s, or even in the 1980s, but underlying structures of inequality are extremely resistant to change, and smallholders remain subject to high levels of vulnerability, exploitation, and social exclusion. We want to conclude by reminding readers that these are not abstract academic concepts but rather conditions that shape the everyday lives, experiences, hopes, and expectations of our fellow human beings. They are very, very real—the conditions that give meaning to work, school, family, childhood, identity, and sometimes even survival. Consider these two reflections by Paraguayan small farmers.

> *We grew corn, cotton, and none of the crops have a good price. We sold them and were left without a cent. We hardly had enough to get through the day, and there wasn't enough for the kids to go to school. And our kids left three years ago to look for work in Argentina, to do any job that is available. I want my kids to come back and work by my side. Even though we don't have much land, our*

work with the cooperative is allowing me to keep putting food on the table. My kids are now back and if our work with the cooperative continues to succeed, it is our hope that the kids will be able to stay. [Farmer 1]

The Brazilians came in, and they're buying up a lot of land, so we Paraguayans are fighting not to lose what we have so we don't suffer hardships. When someone sells the land and gets some money, the money doesn't stretch, the money runs out and the family becomes poor. That's why everyone has to think, has to dialogue with each other to defend our position together as small farmers, so that no one has to sell their property. [Farmer 2]

We compiled this volume to explore the hypothesis that for the most marginalized smallholders in Latin America, cooperatives can become agents of change and mechanisms of equitable and sustainable development. Can cooperatives fulfill these farmers' hopes? Can they not only enhance farmers' incomes but also help smallholders keep their land, sustain their way of life, develop power within commodity chains, and forge meaningful solidarity? Can they help reverse the various axes of powerlessness that make farmers so vulnerable and that permit their ongoing exploitation?

The study of smallholder cooperatives is not new by any means; it has passed through different periods related to both academic and political trends. During the 1960s and '70s, attention to cooperatives flourished as left-leaning intellectuals, activists, and rural leaders hoped cooperatives might transform society by spreading values of equality, solidarity, and participation. The idea was that cooperatives, in the words of Orlando Fals Borda (1971, 146), could be envisioned as "active peasant organizations that would challenge the status quo" by supporting larger social movements that promoted the interests of the peasant class. His research on Latin American smallholder cooperatives, however, led him to the conclusion that cooperatives did not live up to the expectations of their champions. At best, cooperatives served as excuses for inaction rather than as instruments of socioeconomic change. At worst, they were co-opted and eventually reinforced inequality and historical power relations. During the 1980s and '90s, structural adjustment reforms and an emphasis on free markets and the privatization of productive and financial sectors as part of neoliberal globalization became the trend throughout the region. As states withdrew support from smallholder cooperative organizations and accused them of inefficiency, academics and activists lost interest in what appeared to be a largely failed undertaking.

In the wake of the neoliberal wave of the 1990s, smallholder cooperatives persist and may have become more relevant than ever. Indeed, several of the case studies discussed here show promise as palliatives to the poverty, inequality, environmental decline, and record rural outmigration sparked by neoliberal policies. At the same time, these cases reveal the varied and contradictory nature of neoliberalism by illustrating how states continue to provide vital public services, technical assistance, and financing that the private sector would never have deemed profitable. Perhaps evaluations of neoliberalism must move beyond simple dichotomies of the state versus society/corporations to consider the diverse ways that state institutions and grassroots organizations interrelate.

This complexity notwithstanding, we would argue that the political and intellectual movement for more equitable societies must remain highly skeptical of neoliberal globalization while seeking alternatives with global relevance. The lesson from the past is that smallholder cooperatives are not social movements in and of themselves, and they do not necessarily guarantee justice or equity. As state-sanctioned organizations, they must work within the system that they seek to transform, and as such they face enormous contradictions. Cooperatives are, however, increasingly part of a larger global movement that focuses on alternative development (including, for example, the solidarity economy, fair trade, organic farming, land rights movements, and efforts to change global trade rules) and that addresses food security, environmental sustainability, and long-term quality-of-life concerns for rural populations. One of the interesting common threads among the cases presented here is that these cooperatives are significantly transnational and part of a global movement that seeks to integrate historically marginalized smallholders into global markets, global concerns, and power structures that go beyond communities. In several of these case studies we find clear examples of how such transnational work can transform local structures of oppression.

It is also important to highlight that these cooperatives are not linear organizations; instead they have complex histories. Most have undergone periods of crisis and have had the tenacity to develop change strategies that have allowed them to persist. The case studies constitute a wide range of experiences that provide to different degrees a lens through which one can envision smallholder cooperatives as organizations capable of building more autonomous communities that strive to free themselves from historical patterns of exploitation. Thus, the larger question that we ask is, under what circumstances and through what actions do institutions like cooperatives contribute to smallholders' access to, on the one hand, more productive, sustainable, and satisfying livelihoods, and, on the other hand, a genuine forum through which they can participate and engage politically in the transformation of their own communities? The International Cooperative Alliance has tried to stress "the cooperative difference" as an economic selling point, but what is "the cooperative difference" from the perspective of power? What lessons can we draw from these case studies about the conditions, dynamics, and functional characteristics that make cooperatives useful, active, adaptable, and responsive?

Even though we have found a great deal of diversity among the cooperatives that we have examined, each provides valuable lessons and makes an important contribution to our current understanding of smallholder agricultural cooperatives in Latin America. Reading across cases, we draw a number of conclusions and highlight recurring themes.

ECONOMIC GAINS MUST BE TRANSLATED INTO SOCIAL INVESTMENTS THAT SUPPORT POLITICAL-ECONOMIC TRANSFORMATION

In most cases, smallholder cooperative members are multiply marginalized: they have little formal education and few productive resources, and they are embedded in long-standing relations of inequality and exploitation with local power elites. This situation of vulnerability is exacerbated in the case of women, whose potential and needs are frequently ignored in marginal rural settings and who are often overburdened by multiple responsibilities. In this context, cooperatives can play an important role by investing in human capital, empowering producers (men and women), and strengthening the long-term sustainability of peasant agriculture in order to decrease the vulnerabilities that reproduce dependency. The Manduvirá cooperative in Paraguay stands out for its ability to balance economic and social priorities. Despite its tremendous business success, the cooperative has continually invested in its membership by providing educational opportunities for cooperative members and their children and by employing young adults and educating women to occupy leadership positions in the cooperative. Participation in fair trade has also generated the funds to provide health care to the community and has allowed the cooperative to build strategic external linkages that support the cooperative's constant search for autonomy. After the cooperative declared bankruptcy, Manduvirá was able to flourish by leading producers in a common struggle against exploitation by the local sugar mill. Importantly, the economics of cooperativism—the pooling of resources and ownership for the benefit of all—were essential for peasant political mobilization that could successfully change long-standing relations of inequality. The cooperative continues to flourish because it has been able to turn economic gains into human and social capital, impacting not only cooperative members but the entire community of Arroyos y Esteros.

On the opposite end of the spectrum is the Brazilian cooperative Amazon-Coop, which failed to translate economic gains into sustained investments in human capital and social or political-economic change. Indigenous members of this cooperative were positioned as wards rather than as owners of the cooperative, and the model that predominated was highly paternalistic. Cooperative members belonged to different ethnic groups and were located at great distances from one another, and the leadership had little contact with the membership. Even though members were obtaining significantly higher prices for Brazil nuts, the cooperative failed to act as a change agent that would improve indigenous people's position within local power relations. Instead, it re-created historical relations of control and exploitation. It is not surprising that this cooperative eventually collapsed, leaving its members as marginalized as they had been when the cooperative started.

The structural marginalization of Latin American peasants, however, is not a result only of rural class relationships. There is strong evidence from these case studies that democratic cooperatives able to connect popular concerns with management expertise are in a good position to identify and address local sources of vulnerability and marginalization. For example, the Japanese Brazilian cooperative CAMTA and the Paraguayan cooperative of Capiibary both found that environmental degradation was central to local livelihood challenges, and in the case of Capiibary, these paired challenges were prompting widespread land sales, a rural exodus of young farmers, and land concentration by agro-industrial soy farmers. Both cooperatives invested in education, production, and marketing diversification through agroforestry programs, which brought economic stability while addressing soil degradation and deforestation. CAMTA also strengthened cultural cohesion among Japanese Brazilians by linking the cooperative to Japanese markets and financial capital. In

Capiibary, the cooperative has helped reverse the rural exodus that threatened to destroy the community by regenerating farmland, making farming more economically viable, paying attention to all phases and types of agricultural production including those where women have a predominant role, and restoring dignity to peasant life.

In some of the cooperatives, the attention paid to programs that benefit women has been critical and is reflected in the importance given to the participation of women as both cooperative leaders and rural producers. In cases where women are included in the cooperative's strategy of rural development, there is an increase in adaptive capacity not only for women themselves but for the entire household. In the case of coffee associations in Colombia, for example, the inclusion of women has been key in the successful transition from illegal crop production, carried out mostly by men, to household participation in the specialty coffee value chain. In the savings and credit cooperatives in Colombia's southern Santander, the growing participation of women in managerial positions has been fundamental to establishing an inclusionary strategy determined by collective concerns.

The savings and credit cooperatives in Colombia's southern Santander region offer another example of adapting cooperatives to local needs. These cooperatives have invested in the technological development of members' farms to improve business productivity while simultaneously collaborating with local organizations to create "peace territories" and dialogue with armed actors to devise conflict resolution strategies. This is absolutely essential in Colombia, where rural violence has fueled an agrarian counterreform via the direct displacement of more than 4 million people and the indirect displacement of many more who can no longer tolerate the fear and economic instability generated by armed actors. Given the absence of the Colombian state in the south of Santander, cooperatives have played a critical role in building community identity and cohesion based on principles of trust and solidarity, which ultimately ensures that the human rights of their members and the community at large are respected.

These cases illustrate that a key task for cooperatives is not to follow expert models of cooperative success, but rather to develop strategies grounded in local conditions. Building on strong community support and democratic involvement, cooperatives can develop creative strategies that mobilize profits for the transformation of rural inequality and marginalization. Overcoming the formidable obstacles to change, however, requires deep solidarity, trust, and strong cooperative economies.

THE ROLE OF THE STATE IS CRITICAL IN INSTITUTING SUPPORTIVE POLICIES AND PROMOTING AN ENVIRONMENT OF SOLIDARITY

All of the cooperatives in this volume have struggled with national policies that either directly or indirectly affect cooperative decision making, management, investments, and outcomes. To some extent, these cases therefore illustrate the importance of a supportive policy environment for cooperatives. In some cases, cooperatives have benefited directly from government investments, as in the federally funded school lunch program executed by the Manaus Secretariat of Education, which purchases pineapple and cupuaçu from ASCOPE. Similar policies could provide significant support for smallholders across the continent and might be specifically tailored to promote food sovereignty and sustainable farming methods. In other cases, policies contradict official government efforts to promote cooperatives. A case in point is the disastrous impact of the 1999 Paraguayan debt forgiveness law on the Coronel Oviedo cooperative, which illustrates how policies that may be favorable from a household perspective can backfire when they damage community institutions. When the law was issued, the cooperative had already paid its public debt and as a result was not obliged to forgive members' debts. Even though the cooperative offered to renegotiate the terms of loans, it lost more than half of its rural members. Similarly, COLANTA's struggles with Colombian dairy policy and free trade agreements show that even the largest and most lucrative cooperatives may be challenged by global economic policies. One key lesson, then, is that cooperatives must not only respond effectively to state-based challenges and opportunities, but must also work diligently to create and maintain favorable policy conditions (see also Burke and Piekielek 2011).

A second aspect of state influence has to do with effects on cooperative solidarity, and this suggests a more inward-looking strategy for cooperative managers. In Paraguay and Brazil, and to a lesser extent in Colombia, cooperative formation was the direct result of government campaigns to expand the agricultural frontier during the 1970s. Cooperatives offered a way to cope with the absence of government services for credit, production assistance, marketing, and infrastructure. Strong grassroots cooperatives like CAMTA and ASCOPE assumed many of the roles of local government, providing basic services such as electricity in the case of CAMTA, and becoming the central vehicle for community-wide action in the case of ASCOPE. Not surprisingly, abandonment by the state has often strengthened community organization, while community organizing has often facilitated more effective state support. ASCOPE's success, for example, is largely a function of its ability to maneuver effectively within the Brazilian policy environment.

By contrast, cooperatives that have been the direct result of state intervention and have been unable to develop a structure of democratic decision-making tend to be the most fragile cooperatives and the ones that lack a strong sense of solidarity. In Brazil, AmazonCoop and Produtores do Curupati-Peixe (CPCP), the tilapia cooperative, exemplify this. The former was initiated and led by the Brazilian state and a transnational corporation, and the latter started after forced relocation of the local community following the construction of the Castanhão Dam. While the significant economic success at CPCP should not be downplayed, many of the cooperative's members focused on their personal incomes and struggled to appreciate that they were owners of the cooperative enterprise. This may significantly limit the cooperative's ability to more broadly transform rural inequality.

Of course, dependence need not be a permanent condition. In Paraguay, Guayaibi Unido has been highly dependent on state programs, and its development strategy is largely dictated by CEPACOOP. However, this cooperative took an important step in building solidarity when it was able to use mass mobilization to challenge the lack of government support when their banana shipments were stopped at the Argentina-Paraguay border. Although the Manduvirá cooperative has been more successful in its use of strikes and mass mobilization, both cooperatives have challenged the political and social system to further the interests of their members and communities. By publicly challenging conditions that threaten members' subsistence, cooperatives can facilitate participation in larger movements that seek structural change and can give members a deeper sense of empowerment and agency.

Importantly, our cases show that it is not only abandonment and opposition that give rise to solidarity. Cuéllar and Ramirez discuss how Colombian coffee-producer organizations were able to leverage strong support from USAID (via

ACDI/VOCA) and the Colombian government in order to stabilize coffee farmers' livelihoods and establish stronger democratic organizations. Even they struggled with conflictual and changing policies, however.

CHANGE STRATEGIES MUST BE TAILORED TO (AND OFTEN MUST CHALLENGE) LOCAL CULTURAL PATTERNS

Smallholder cooperatives in Latin America, particularly in marginal regions, develop in a cultural and political context that is marked by social stratification embodied in systems of patronage and clientelism. Within this context, cooperatives may be dominated by local elites or more powerful external agents, and grassroots actors may actually contribute to this domination by pursuing opportunities offered within the patronage systems. The case of AmazonCoop provides, again, an example of an extreme case. The stimulus that created this cooperative first came from public officials who felt frustrated at the inability of the public sector to meet the needs of their indigenous clientele. But just as larger policy toward indigenous peoples in Brazil tends to be highly paternalistic, so has been the evolution of AmazonCoop in relation to its indigenous membership. In many ways, AmazonCoop reflects the larger institutional and legal frameworks that define indigenous-Brazilian relationships.

Other cooperatives, such as Guayaibi Unido, are inadvertently reproducing a pattern of social stratification where the more disadvantaged banana producers retreat from active participation while top producers continue to grow through their linkages to the central cooperative, which plays a critical role in determining who benefits from the export market. In this cooperative, management success requires a great deal of work in the empowerment of all cooperative members and the just distribution of benefits.

Some cooperatives, like Manduvirá, have made a conscious effort to end patterns of exploitation based on social class by directly challenging the power of local elites involved in the buying of sugarcane. Manduvirá's unprecedented success has resulted in a membership that feels empowered and capable of challenging one of the largest industrial sugarcane producers in Paraguay. ASCOPE also offers an interesting example, where a clear ideology of class consciousness and social equality has provided a foundation on which the cooperative could develop a successful market-based economic model. Its vision, however, is economic success without *ever* re-creating old patterns of exploitation and social inequality.

Taken together, these cases illustrate the importance of understanding historical power structures and the ways that they shape both economic processes *and* cultural expectations about what types of changes are possible and desirable. To be agents of change, cooperatives need to accommodate their work to local cultural expectations while simultaneously challenging those cultural dynamics that reproduce disempowerment and marginalization.

SUCCESSFUL PARTICIPATION IN INTERNATIONAL MARKETS REQUIRES A FLEXIBLE STRATEGY TO AVOID OVERDEPENDENCE ON THOSE MARKETS

The prevalent trend among the cooperatives studied is to gear production more toward international markets through the creation of strategic alliances that permit cooperatives to participate in niche markets, including organic-certified production and fair trade. This trend is itself a powerful transformation of rural political economies: it signifies a move away from reliance on local and regional intermediaries and toward vertical integration and industrial processing to keep more profits within the cooperative. As such, it enables small and medium-size farmers to seize global market opportunities that, were they to operate as individuals, would be beyond their investment capabilities and skill sets. Without question, this type of international strategy presents a potential growth area for rural cooperatives. It stimulates the development of new kinds of collaboration among actors along commodity networks, giving rise to new levels of organization with important implications for reducing rural poverty. It may also have positive multiplier effects across the rural economy by providing local sources of off-farm income, as shown by the Colombian dairy cooperative COLANTA and the Colombian rural savings and credit cooperatives.

However, because the mission of cooperatives goes beyond the provision of material benefits, the international strategy may also limit cooperatives' ability to fulfill principles of democracy, participation, autonomy, and self-sufficiency. Those cooperatives that participate in international marketing networks must conform to transnational regulatory frameworks and quality standards that require increasing production costs and reorganizing cooperatives, with no long-term guarantee of more favorable trading conditions. Market and product diversification is a key strategy to achieve independence and reduce the risk of overreliance on a single marketing network or product.

CAMTA, Capiibary, Manduvirá, and the Colombian specialty coffee associations have successfully entered commodity markets; the first two have worked actively to avoid excessive dependence upon those markets. CAMTA's long experience in international commodity markets has led it to emphasize a strategy of diversification and flexibility. To spread risk, CAMTA established a mixed agroforestry model that combines marketable native fruits with introduced tropical fruits and native timber species. To reduce dependence on uncertain markets, CAMTA has established an agro-industrial processing capacity. In the case of Capiibary, the cooperative has developed a multipronged strategy that has allowed smallholders to participate in markets from which they have traditionally been excluded, like soy, and to partner with private companies to process their fruit for sale in national and international markets. At Manduvirá, leaders have made substantial efforts to educate themselves in fair trade marketing. Even though Manduvirá is becoming increasingly independent as it starts its own sugar processing, its strategy of dependence on sugar does not have the element of diversification that is critical to creating sustainability. A challenge that the cooperative faces is to be able to manage the cooperative's involvement in certified-trade networks rather than finding its internal participatory processes dictated by regulatory and economic dynamics tied to fair trade and certification. It is clear that the increasing opportunities opened by marketing networks are requiring farmers to become highly specialized in terms of agricultural production, marketing, business, and management. The leaders of all of these cooperatives and associations have been able to mobilize members, create competition, and change the balance of power. They have been able to take advantage of the opportunities of globalization, but some remain dangerously dependent on those opportunities.

Neoliberal policies of free trade and the establishment of MERCOSUR have had positive and negative impacts on cooperatives in all three countries that we consider here. For example, in Paraguay, Guayaibi Unido's alliance with CEPACOOP has permitted higher earnings through banana exports to Argentina, but at the cost of subjecting farmers to more volatile markets and competition with more technified producers. With adequate support, Guayaibi Unido has the potential to follow a similar path as the above cooperatives, but this requires closing the gap between those who are producing export-quality bananas and those who feel left out when their bananas are not considered good enough. In other places—such as the Brazilian and Colombian dairy sector—the gap between Latin American farmers and large-scale, highly subsidized European and North American farmers may simply be too large to close absent significant changes to free trade agreements.

COOPERATIVES MUST ACHIEVE EFFECTIVE LEADERSHIP WITH STRONG PARTICIPATORY GOVERNANCE

Developing and maintaining a sense of ownership, collective identity, and empowerment among the membership while promoting an entrepreneurial culture requires the ability to balance external linkages with internal governance. In cooperatives that lack a sense of ownership, leaders and members often make a strong us/them distinction. Sense of ownership is related to the historical context in which each particular cooperative developed, as well as to a cooperative's degree of heterogeneity and size. In cases where the cooperative is imposed by political bosses, external agents, or the state, the sense of ownership tends to be lower. In cases where there is great socioeconomic differentiation among members in terms of class, wealth, ethnicity, or access to resources, the sense of ownership tends to be lower. That said, a heterogeneous membership can be a useful political and economic resource for cooperatives.

In the Paraguayan cooperative Coronel Oviedo, the size of the membership and its dispersal across the agricultural landscape have contributed to a sense of separation between management and producers. Rural members tend to feel intimidated and unwilling to participate and have little sense of loyalty to the cooperative. This is related to the historical emphasis placed on the urban sector, which is in turn related to the fact that the board of directors is entirely composed of urban members. However, the mix of urban and rural members creates opportunities for important economic linkages and new social programs, and urban members often subsidize rural ones.

Cooperatives with a high sense of ownership stand in contrast. A particular case is CAMTA, the Japanese Brazilian cooperative. A strong sense of ownership in this case can be partly attributed to the ethnic homogeneity underlying cooperative membership and a historical sense of common struggle. This cooperative has been able to effectively deal with crises because of strong loyalty from members, who contributed their personal resources to keep the cooperative solvent during a difficult period. The case of ASCOPE is also revealing. A strong sense of ownership is partly related to a common historical struggle for land and a long experience of grassroots mobilization, social consciousness, and autonomy. The cooperative's deep ties to the community are also critical, as is the small size of the group and the physical proximity of residents whose design of the agro-villa model allows everyone to access basic services.

Nurturing a sense of ownership must also be accompanied by a cooperative management with knowledge of good business practices that promotes an entrepreneurial culture. The organization and professionalism in cooperative management varies a great deal among the cooperatives studied. As in other realms of cooperative organization where a balance between objectives does not intuitively coexist, we find that successful internal governance requires the ability to accommodate the interests of sometimes highly diverse members while at the same time developing certain standard procedures and ways of categorizing members that allow for the development of a more homogeneous group that is easier to manage. This is a particular challenge for cooperatives that are large and whose members are widely dispersed, such as Coronel Oviedo, where, as already stated, members do not have a strong sense of sharing power in the decision-making process. Good governance may also require difficult decisions, as in the case of the Colombian coffee associations that had to purge members in order to strengthen the cooperative spirit.

For small cooperatives such as Guayaibi Unido and ASCOPE, the problems of leadership are different. In these two cases, the role of director is not distinguished from that of manager, and leadership is more related to charisma than professional management skills. This leads to a concentration of decision-making power and a general sense of lack of transparency and accountability, particularly as these cooperatives adopt more business-oriented models and expand economic activities. Other cooperatives are struggling to professionalize their management and maintain well-defined leadership roles in order to assure transparency. Such is the case at Manduvirá. As it expands its activities and increases in size, there is a clear awareness that it needs to maintain a group of professionals that can tackle different aspects of cooperative organization, from computer-based accounting systems to an understanding of world markets and prices.

CAMTA also offers an interesting example of how the cooperative can maintain a vital organization. In this case, the "old guard" of Japan-born leaders has been replaced by Japanese Brazilians. This cultural and generational shift in leadership contributes to the dynamism of the cooperative and its ability to adapt and pursue a sustainable expansion strategy.

The area of planning also relates to organizational management. In this regard, we found few instances of long-term strategic planning and very little evidence of monitoring and evaluation of the cooperatives and of short-term plans. Even though each cooperative has a vision and some plans to accomplish this vision, these are not well-defined or detailed. Strategic planning requires the specific definition of agro-industrial projects, including the identification of potential difficulties. It also requires a clearly defined path of agricultural outreach and the identification of potential markets and activities. Thus, for cooperatives to achieve organizational sustainability, a blend of stability, flexibility, and forward planning is required, and the cooperative's leadership must ensure that their organizational structure provides these three elements.

ANTICIPATING FUTURE CHALLENGES AND OPPORTUNITIES

Moving forward, some of the most important work for scholars and cooperative managers will involve analyzing the social, political, and economic dynamics that will impinge on cooperative success, and strategizing around the challenges and opportunities of the future. Here we would like to highlight three of these. First, there is the on-going leftist shift in regional politics—the so-called "pink tide." This raises valuable opportunities for scholars to interrogate the extent to which leftist/populist politics generate favorable conditions for cooperatives, and especially to analyze what series of conditions may lead paradoxically to *un*favorable conditions. For cooperative leaders, the rise of left-leaning governments and more democratic populism creates a number of political openings. For example, anti-imperialist and antiglobalist ideologies dovetail nicely with food sovereignty initiatives and crop diversification programs that might reduce vulnerability to international markets; popular discourses supporting socialism and socially just policies could reinvigorate movements for the more equitable redistribution of land and water; and populist appeals could be used to rally support around grassroots cooperatives as part of the people's economy. A great deal of political work will be necessary to ensure that these leftist and populist discourses generate truly transformative policies—and current events are not promising in this regard—but if cooperatives are able to mobilize members, then they may be especially powerful change agents in this regard.

Second, there is the lurking danger of climate change and changing energy economies. Climate change is already challenging Latin American farmers, and it is likely to have a disproportionate impact on the most vulnerable. Because climate acts as a "threat multiplier" that exacerbates other vulnerabilities (Crate and Nuttall 2009; Eakin 2006; Vásquez-León 2009), any cooperatives that help to stabilize rural livelihoods and reduce sociopolitical exclusion will have a buffering effect. Cooperatives may also directly contribute to climate resilience by integrating "climate-smart" technologies and practices into their current mix of technical assistance, and by maintaining the strong social networks and community institutions that are necessary for resilience. Savvy cooperatives may be able to take advantage of new funding opportunities in this regard. However, there are many more complex and uncertain aspects to the relationship between cooperativism, climate, and energy. For example, any effective climate change mitigation strategy will require leaving oil in the ground. Assuming that countries of the global North do not bear full responsibility for this, then some Latin American countries will need to find alternative ways of ensuring citizens' well-being and economic stability in the absence of oil revenues. Rural cooperatives may be a key tool for this post-oil transition. Similarly, meaningful climate policy will necessarily affect fuel prices and global transportation costs, with uncertain effects on international trade and export-oriented cooperative strategies. Cooperative scholars and leaders will need to keep a careful eye on these changes.

Third, land grabbing is becoming widespread. In recent decades, Latin America, like other areas of the developing world, has seen a massive and rapid wave of land acquisition by national elites and foreign governments and corporations. Driven by food, energy, climate, conservation, and financial crises—and by the investment opportunities that these crises present—land grabbing is "likely to result in widespread reconcentration of land and capital" (Borras et al. 2012, 867). Land grabbing thus exacerbates the already-prevalent agrarian counterreform prompted by neoliberal economic globalization, which has tended to exacerbate rural inequalities by creating "an agriculture of two velocities," as capitalist farmers have the capital, technical knowledge, economies of scale, and marketing skills to benefit from new export markets while peasant producers suffer from the lack of those assets and opportunities, as well as from the withdrawal of state supports and increased competition from cheap, subsidized food imports (Kay 2008, 918; Kay 2006). In this context, even policies that could favor smallholders—such as the formalization of land tenure arrangements and property markets—have often facilitated the loss of their most basic resource: land. Many of the cooperatives we have studied show real promise of helping campesinos maintain access to land. They have also helped campesinos resist or even counter key aspects of "new ruralization," including semi-proletarianization, the feminization of rural work, and out-migration (Kay 2008). Cooperatives will need to redouble their efforts to support campesino livelihoods and long-term resource access as market pressures on land, water, and subsoil minerals increase.

CONCLUDING THOUGHTS

Cooperatives are no more a panacea today than they were in the 1970s, when engaged scholars such as Orlando Fals Borda and June Nash were investigating their potential for ending the long-standing exploitation of the Latin American peasantry. However, they may be more important than ever given growing inequality in rural Latin America, the associated exclusion of smallholders from new economic opportunities, and neoliberal policies that withdraw state supports while exposing farmers to fierce market competition. Indeed, the case studies collected here show that successful cooperatives serve as powerful change agents, offering short-term economic and social benefits to member households *and* helping to counter the long-term processes through which exploitation, marginalization, and vulnerability are reproduced. By making visible their separate struggles, we have sought to reveal useful lessons about how to empower the most marginalized sectors of contemporary Latin America via the simultaneous promotion of equitable development and effective collective action. We cannot stress enough that these goals must go hand in hand, that projects promoting development and well-being must be linked to movements for justice and transformation. And while we have certainly witnessed our share of failures—cooperatives that were woefully mismanaged, or that never achieved financial solvency, or that were co-opted by the state or local elites—we have concluded that cooperatives are among the institutions most capable of advancing these twin goals of development and justice. Finally, we hope that by drawing attention to cooperatives' economic and social goals, their participatory and democratic ideals, their strong orientation toward collective action, and their progressive notions of ownership and grassroots empowerment, this volume will encourage readers to seek new, creative, and concrete ways to advance the struggle for a just, equitable, and sustainable world.

REFERENCES

Abramson, Evan. 2009. "Soy: A Hunger for Land." *NACLA Report on the Americas*. May/June.

Ariza, Leonardo, and Luis Lobo. 2002. "Cooperativismo y Bienestar: El Caso de COAGRANJA Ltda." In *Gestión y Desarrollo: La Experiencia de las Cooperativas en Colombia*, eds. Juan Manuel Silva and Ricardo Dávila, 297–358. Bogotá: Pontificia Universidad Javeriana, Facultad de Estudios Ambientales y Rurales.

Ariza Ramírez, Leonardo Alberto. 2007. "Plan de Fortalecimiento Cooperativo y su Impacto en el Desarrollo Rural: Casos de COAGRANJA y COOPVALLE Ltda." In *La Intercooperación del Concepto a la Práctica*, ed. Margarita Oseguera de Ochoa, 141–54. Quebec: IRECUS, Université de Sherbrooke.

Ballvé, Teo. 2013. "Grassroots Masquerades: Development, Paramilitaries, and Land Laundering in Colombia." *Geoforum* 50: 62–75.

Banco de la República. 2011. "Division Política Administrativa de Colombia." Retrieved from http://www.banrepcultural.org/blaavirtual/revistas/credencial/enero2002/division.htm. Accessed May 26, 2011.

Bareiro, Victor. 1997. *Notas sobre el Proceso Agrario Paraguayo*. Pastoral Social Nacional, Servicio Jurídico Integral para el Desarrollo Agrario. Asunción: El Lector.

Barham, Bradford, and Oliver Coomes. 1994. "Wild Rubber: Industrial Organisation and the Microeconomics of Extraction during the Amazon Rubber Boom (1860–1920)." *Journal of Latin American Studies* 26(1): 37–72.

Bennett, John W. 1983. "Agricultural Cooperatives in the Development Process:

Perspectives from Social Science." *Studies in Comparative International Development* 18(1–2): 3–68.

Berdegué, Julio A., and Ricardo Fuentealba. 2011. "Latin America: The State of Smallholders in Agriculture." Paper presented at the IFAD Conference on New Directions for Smallholder Agiculture. January 24–25. IFAD, Rome.

Bialoskorski, Sigismundo. 2004. *Economia das Organizações Cooperativas: Uma análise da influencia da cultura e das instituições.* Tese Livre Docência, Universidade de São Paulo.

———. 2006. *Aspectos Econômicos das Cooperativas.* Belo Horizonte: Mandamentos.

Bolfe, Edson Luis and Mateus Batistella. 2011. "Análise Florística E Estrutural de Sistemas Silviagrícolas Em Tomé-Açu, Pará." *Pesquisa Agropecuária Brasileira* 46: 1139–147.

Borras, Saturnino M., Jr., Jennifer C. Franco, Sergio Gómez, Cristóbal Kay, and Max Spoor. 2012. "Land Grabbing in Latin America and the Caribbean." *Journal of Peasant Studies* 39(3–4): 845–72.

Bucheli, Marieta. 2002. "El Proceso de Crecimiento y Permanencia de una Cooperativa Rural en Colombia: El Caso de la Cooperativa de Ahorro y Crédito para el Desarrollo Solidario de Colombia, COOMULDESA." In *Gestión y Desarrollo: La Experiencia de las Cooperativas en Colombia*, eds. Juan Manuel Silva and Ricardo Dávila, 261–96. Bogotá: Pontificia Universidad Javeriana, Facultad de Estudios Ambientales y Rurales.

———. 2006. *Curas, Campesinos y Laicos como Gerentes Del Desarrollo: La Construcción de un Modelo de Desarrollo Emergente en Colombia.* San Gil, Santander, Colombia: EDISOCIAL.

Bunker, Stephen G. 1984. "Modes of Extraction, Unequal Exchange, and the Progressive Underdevelopment of an Extreme Periphery: The Brazilian Amazon, 1600–1800." *The American Journal of Sociology* 89(5): 1017–64.

Burke, Brian J. 2006. *Indigenous Cooperatives, Corporations, and the State on Brazil's Extractive Frontier: Contemporary and Historical Globalizations.* Master's thesis, Department of Anthropology, University of Arizona.

———. 2010. "Cooperatives for 'Fair Globalization'? Indigenous People, Cooperatives, and Corporate Social Responsibility in the Brazilian Amazon." *Latin American Perspectives* 37(6): 30–52.

Burke, Brian J., and Jessica Piekielek. 2011. "Cooperatives, Politics, and Development in Rural Paraguay." *Human Organization* 70(4): 355–65.

Burke, Brian J., and Paola Canova. 2006. "Change Strategies of Latin American Cooperatives: The Case of CAMPEALTA/Amazon Coop (Altamira, Brazil)."

Report for Agricultural Cooperative Development International/Volunteers in Overseas Cooperative Assistance. Tucson: BARA.

Campos, David. 2003. *Impacto de la Estrategia de Desarrollo Rural Humano Agroecológico en la Reducción de la Pobreza en las Comunidades Campesinas de Loreto y RI3 Corrales en Paraguay*. Asunción: Sociedad de Estudios Rurales y Cultura Popular.

Carosini, Leticia. 2012. *Vision Panoramica del Sector Cooperativo en Paraguay: Una Importante Contribución al Desarrollo Nacional*. La Paz, Bolivia: Organizacion Internacional del Trabajo, Oficina para los Paises Andínos.

Carter, Michael R., Bradford L. Barham, and Dian Mesbah. 1999. "Agricultural Export Booms and the Rural Poor in Chile, Guatemala, and Paraguay." *Latin American Research Review* 31(1): 33–65.

Casa de Cultura. n.d. *V Departamento de Caaguazú: Cifras, Referencias, y Estadísticas*. Coronel Oviedo: Casa de Cultura.

Castillo, Darío, Paulo Albuquerque, Jorge Bertullo, Juan José Sarachu, Diego Barrios, Marcela Setaro, Mario Radrigán, Rodrigo Flores, Karin Baeza, and Nicolás Montoya. 2005. "Desarrollo Cooperativo, Políticas Públicas e Integración Regional. Estudio de Caso en el Contexto del Mercado Común del Sur, MERCOSUR." *Unircoop* 3(1): 9–31.

Castro, Eduardo Viveiros de. 1992. *Araweté: O Povo do Ipixuna*. São Paulo: Centro Ecumênico de Documentação e Informação.

Cohn, Patricia J., Matthew S. Carroll, and Jo Ellen Force. 2013. "So Happy Together or Better Off Alone? Women's Economic Activities, Cooperative Work, and Empowerment in Rural Paraguay." In *Cooperatives and Community Development*, eds. Vanna Gonzales and Rhonda G. Phillips, 93–109. London: Routledge.

COLANTA. 2010. *Informe de Gestión 2009*. Medellín: COLANTA.

———. 2011. "Tiempos Críticos: Importaciones, Enlechadas." http://sabemas.colanta.com.co/sabemas.net/historia.asp. [Accessed August 12, 2011.]

———. 2012a. "Colanta—Calidad." http://www.colanta.com.co/colanta-institucional/calidad. [Accessed June 6, 2012.]

———. 2012b. "Colanta—Nuestra Historia." http://www.COLANTA.com.co/index.php?option=com_content&task=view&id=17&Itemid=231. [Accessed June 6, 2012.]

Colantapoli. 2009. "Colanta en Cifras." COLANTA blog. May 26, 2009. http://colantapoli.over-blog.es/article-31869059.html. [Accessed June 6, 2012.]

Colombo, Sandra, and Paula Oxoby. 2013. *El Sector Cooperativo en el MERCOSUR Durante la Primera Decada del Siglo XXI: Retos y Oportunidades*. Málaga: Centro de Estudios Interdisciplinarios en Problematicas Internacionales y Locales, Fundación Universitaria Andaluza "Inca Garcilaso."

Comisión de Seguimiento a la Política Pública Sobre Desplazamiento Forzado. 2009. *El Reto ante la Tragedia Humanitaria del Desplazamiento Forzado: Reparar de Manera Integral el Despojo de Tierras y Bienes*. Bogotá: Consultoría de Derechos Humanos.

Confederación de Cooperativas de Colombia—Confecoop. 2010. "Las Tasas de Interés en el Sector Cooperativo Colombiano." *Observatorio Cooperativo* 12. Retrieved from http://www.confecoop.coop/observatorio/12/No.%2012–1.pdf. [Accessed June 17, 2011.]

CONPES (Consejo Nacional de Política Económica y Social). 2010. *Política Nacional para Mejorar la Competitividad del Sector Lácteo Colombiano*. Documento Conpes 3675. Bogotá: Departamento Nacional de Planeación.

Coque Martínez, Jorge. 1996. "El Fomento de Modelos Empresariales Cooperativos como Estrategia de Desarrollo Local en Zonas Rurales: El Caso de la Provincia Guanentá, Departamento de Santander, Colombia." *Cuadernos de Desarrollo Rural* 36: 63–80.

Crate, Susan A., and Mark Nuttall. 2009. "Introduction: Anthropology and Climate Change." In *Anthropology and Climate Change: From Encounters to Actions*, eds. Susan A. Crate and Mark Nuttall, 9–37. Walnut Grove, CA: Left Coast Press.

Dávila, Ricardo. 1982. "Las Cooperativas: Una Salida Económica o una Perspectiva Social para el Desarrollo Rural." *Cuadernos de Agroindustria y Economía Rural* 8. Bogotá: Pontificia Universidad Javeriana.

———. 1996. "Las Cooperativas Rurales y el Desarrollo Regional: El Caso de las Provincias del Sur de Santander, Colombia." *Cuadernos de Desarrollo Rural* 37: 47–62.

———. 2002. "COMULSEB, Una Alternativa para el Desarrollo Local." In *Gestión y Desarrollo: La Experiencia de las Cooperativas en Colombia*, eds. Juan Manuel Silva and Ricardo Dávila, 215–60. Bogotá: Pontificia Universidad Javeriana, Facultad de Estudios Ambientales y Rurales.

———. 2004. *Innovación y Éxito en la Gerencia Cooperativa: Casos Exitosos de Cooperativas Rurales de Ahorro y Crédito*. Bogotá, Colombia: Pontificia Universidad Javeriana, Facultad de Estudios Ambientales y Rurales.

———. 2008. "La Experiencia Cooperativa del Sur de Santander: Las Cooperativas Rurales de Ahorro y Crédito." In *Desarrollo Económico Local: Una Apuesta por el Futuro*, ed. Luz Stella Sierra de Arango, 296–319. Bogotá: Fundación Social.

Dávila, Ricardo, and Marieta Bucheli. 1998. "Las Cooperativas Rurales y la Crisis del Sector Financiero Cooperativo: El Caso de Colombia." In *Memorias III: Seminario Internacional "Las Cooperativas Rurales y el Desarrollo Regional,"* 25–43. Municipío de Páramo, Santander, Colombia.

Dávila, Ricardo, Miguel Arturo Fajardo Rojas, Oscar Bastides Delgado, Felipe Ortega, and Guiseppina Daros. 2005. *Éxito e Innovación en la Gestión: Las Cooperativas como Agentes de Desarrollo Local.* Bogotá: UNIRCOOP Andino/Pontificia Universidad Javeriana/UNISANGIL.

Dávila, Ricardo, and Jaime Forero. 1987. "Las Cooperativas Rurales de Santander: Conclusiones de un Trabajo de Investigación Participativa." *Cuadernos de Desarrollo Rural* 18: 21–43.

———. 1997. "Cooperativismo y Desarrollo Rural en una Provincia Colombiana." *Cuaderos de Desarrollo Rural* 38–39: 25–33.

Dávila, Ricardo, and Juan Silva. 1996. "Una Financiera Cooperativa en la Provincia." *Cuadernos de Administración* 19: 79–97.

De Sousa Santos, Boaventura, ed. 2006. *Another Production Is Possible: Beyond the Capitalist Canon.* London: Verso.

Deere, Carmen D., Niurka Perez, and Emel Gonzales. 1993. "The View from Below: Cuban Agriculture in the 'Special Period in Peacetime.'" *Journal of Peasant Studies* 21: 194–234.

Dinero. 2009. "Alimentos Funcionales: Crece la Apuesta." *Dinero* 337 (October 30, 2009): 60.

Dirección General de Censos y Estadísticas (DGCE). 2002. *Censo Nacional de Población y Viviendas.* Asunción: DGCE.

Eakin, Hallie. 2006. *Weathering Risk in Rural Mexico: Climatic, Institutional, and Economic Change.* Tucson: University of Arizona Press.

Echeverry Marín, Jaime. 2010. "El Valor de la Democracia." *Así Cumplimos: Periódico Institucional de la Administración San Pedro de los Milagros* 5 (April 2010): 2–3.

Eckstein, S., and T. Wickham-Crowley. 2003. *Struggles for Social Rights in Latin America.* New York: Routledge.

El Colombiano. 2010. "Colanta Ya Puede Exportar a Rusia." *El Colombiano.* March 29, 2010. http://www.elcolombiano.com/BancoConocimiento/C/colanta_ya_puede_exportar_a_rusia/colanta_ya_puede_exportar_a_rusia.asp. [Accessed June 6, 2012.]

El Espectador. 2010. "Colanta, Alpina, Colácteos y Alquería, Autorizadas para Exportar a Chile." *El Espectador* (March 19, 2010): 14. http://www.elespectador.com/economia/articu10193976-COLANTA-alpina-colacteos-y-alqueria-autorizadas-exportar-chile. [Accessed June 6, 2012.]

El Tiempo. 2006. "Los 8.500 Socios Definirán Hoy El Rumbo De Colanta." *El Tiempo.* March 26, 2006. http://www.eltiempo.com/archivo/documento/MAM-1964221. [Accessed June 6, 2012.]

Fairtrade International. 2014. "Paraguayan Pride: Farmers Open World's First Fairtrade Organic Sugar Mill." http://www.fairtrade.net/new/latest-news/single-view/article/paraguayan-pride-farmers-open-worlds-first-fairtrade-organic-sugar-mill.html. [Accessed January 20, 2015.]

Fair Trade Labeling Organization (FLO). 2007. "Sugar." http://www.fairtrade.net/sugar.html. (Accessed May 26, 2007.)

———. 2009. "Fairtrade Standards for Cane Sugar for Small Producers' Organizations." http://www.fairtrade.net/fileadmin/user_upload/content/Cane_Sugar_SPO_Mar_09_EN.pdf.

Fals Borda, Orlando. 1971. *Cooperatives and Rural Development in Latin America: An Analytic Report*. Vol. 3. Geneva: United Nations Research Institute for Social Development.

Fals Borda, Orlando, Raymond Apthorpe, and Inayatullah. 1976. "The Crisis of Rural Cooperatives: Problems in Africa, Asia, and Latin America." In *Popular Participation and Social Change: Cooperatives, Collectives, and Nationalized Industry*, eds. June Nash, Jorge Dandler, and Nicholas S. Hopkins, 439–56. World Anthropology Series. The Hague: Mouton.

Farnan, Tara Colleen. 2007. "Small Farmers and the Agro-Industrial Production and Services Cooperative of Coronel Oviedo: Opportunities for Growth." Master's thesis, Graduate Center for Social and Public Policy, Duquesne University.

Fausto, Carlos. 2001. *Inimigos Fiéis: História, Guerra, e Xamanismo na Amazônia*. São Paulo: Editora da Universidade de São Paulo.

Fernández Valdovino, Carlos G., and Alexander Monge Naranjo. 2004. *Economic Growth in Paraguay*. Economic and Social Study Series. Inter-American Development Bank.

Finan, Timothy J., and Donald R. Nelson. "Making Rain, Making Roads, Making Do: Public and Private Adaptations to Drought in Ceará, Northeast Brazil." *Climate Research* 19: 97–108.

Fisher, William H. 2000. *Rain Forest Exchanges: Industry and Community on an Amazonian Frontier*. Washington, DC: Smithsonian Institute Press.

Floro, Maria S., and Anant Pichetpongsa. 2010. "Gender, Work Intensity, and Well-Being of Thai Home-Based Workers." *Feminist Economics* 16(3): 5–44.

Forero Álvarez, Jaime, ed. 2010. *El Campesino Colombiano: Entre el Protagonismo Económico y el Desconocimiento de la Sociedad*. Bogotá: Pontificia Universidad Javeriana.

Forero Álvarez, Jaime, and Ricardo Dávila Ladrón de Guevara. 1997. "Cooperativismo y Desarrollo Rural en una Provincia Colombiana." *Cuadernos de Desarrollo Rural* 38–39: 25–33.

Formento, Liliana I. 2003. *El Paraguay Campesino: una vieja historia de resistencia, adaptación y funcionalidad.* Río Cuarto, Argentina: Facultad de Ciencieas Humanas, Universidad Nacional de Río Cuarto.

Gagnon, Gabriel. 1976. "Cooperatives, Participation, and Development: Three Failures." In *Popular Participation and Social Change: Cooperatives, Collectives, and Nationalized Industry*, eds. June Nash, Jorge Dandler, and Nicholas S. Hopkins, 365–380. The Hague: Mouton.

Galeano, Luis. 2003. "Movimiento Campesion Hoy: Conquistas y Derrotas en un Contexto Contradictorio de Crisis Social y Política." *Novapolis: Revista de Estudios Políticos Econtemporáneos* 2: 32–38.

Garay Salamanca, Luis Jorge, Fernando Barberi Gómez, and Iván Cardona Landínez. 2009. *Impact of the U.S.–Colombia FTA on the Small Farm Economy in Colombia.* Bogotá: Independent Report Financed by Oxfam America. http://www.usofficeoncolombia.org/uploads/application-pdf/Colombia%20FTA%20impact%20°n%20Small%20Farmers%20-%20Final%20English%20Small.pdf. [Accessed June 6, 2012.]

———. 2010. "Impactos del TLC con Estados Unidos sobre la Economía Campesina en Colombia." In *El Campesino Colombiano: Entre el Protagonismo Económico y el Desconocimiento de la Sociedad*, ed. Jaime Forero Álvarez, 39–56. Bogotá: Pontificia Universidad Javeriana.

GIZ (Deutsche Gesellschaft für Internationale Zusammenarbeit). 2016. "Sustainable Smallholder Agriculture." Programme for Sustainable Natural Resource Management, Paraguay. Ministry of Agriculture and Livestock, Paraguay. https://www.giz.de/en/worldwide/23423.html.

Goodman, David, and Michael Redclift. 1982. *From Peasant to Proletarian: Capitalist Development and Agrarian Transitions.* Oxford: Blackwell.

Gobernación de Santander. 2011. *Plan de Desarrollo Departamental 2008–2011.* http://www.santander.gov.co/pdds/documentos/pdd_santander_incluyente.pdf?phpMyAdmin=5142f4bf7d84972ed3a78acc43f8dc5c. [Accessed June 14, 2011.]

Gonzales, Vanna, and Rhonda G. Phillips. 2013. *Cooperatives and Community Development.* London: Routledge.

González, Andrés Felipe. 2010. "Conpes Lechero No 'Cuajó' entre Productores." *La Nación.* July 22, 2010. http://www.lanacion.com.co/2010/07/22/conpes-lechero-no-%E2%80%98cuajo%E2%80%99-entre-productores/. [Accessed June 6, 2012.]

Grandin, Greg. 2010. *Fordlandia: The Rise and Fall of Henry Ford's Forgotten Jungle City.* London: Macmillan.

Grimes, Kimberly M., and B. Lynne Milgram, eds. 2000. *Artisans and Cooperatives: Developing Alternative Trade for the Global Economy*. Tucson: University of Arizona Press.

Gudoshnikov, Sergei. 2001. "Organic Sugar-Niche Commodity in the Mainstream Market." *FO LITCH International Sugar and Sweetener Report* 133(22): 1–9.

Gutiérrez, Juan David. 2014. "Smallholders' Agricultural Cooperatives in Colombia: Vehicles for Rural Development?" *Desarrollo y Sociedad* 73: 219–271.

Healy, Kevin. 2001. *Llamas, Weavings, and Organic Chocolate: Multicultural Grassroots Development in the Andes and Amazon of Bolivia*. Notre Dame, IN: University of Notre Dame Press.

Hernández Castillo, Rosalva Aída, and Ronald Nigh. 1998. "Global Processes and Local Identity Among Mayan Coffee Growers in Chiapas, Mexico." *American Anthropologist* 100(1): 1–12.

Homma, Alfred Kingo Oyamao. 2004. "Dinâmica dos sistemas agroflorestais: o caso da Colônia Agrícola de Tomé-Açu, Pará." *Revista Instituto de Estudo Superiores da Amazônia* 2(1–2): 57–65.

Homma, Alfred Kingo Oyamao, Antônio José Elias Amorim de Menezes, and Aldecy José Garcia de Moraes. 2014. "Dinâmica Econômica, Tecnologia e Pequena Produção: O Caso da Amazônia." In *O Mundo Rural no Brasil do Século 21*, eds. Antônio Márcio Buainain, Eliseu Alves, José Maria Silveira, and Zander Navarro, 981–1008. Brasília: Embrapa.

IAlimentos. 2009. "COLANTA, Segunda Empresa de Alimentos Más Vendedora del País." Entrevista con el Gerente Jenaro Pérez Gutiérrez. *IAlimentos* (10): 47–51.

IBGE (Instituto Brasileiro de Geografia e Estatística). 2004. *Produção do Extração Vegetal e da Silvicultura*. Vol. 19. Rio de Janeiro: Instituto Brasileiro de Geografia e Estatística.

IICA (Inter-American Institute for Cooperation in Agriculture). 2004. *Caracterización de la Agricultura Familiar Campesina en el Paraguay*. San José, Costa Rica: IICA.

Illovo Sugar. 2007. "World of Sugar." http://www.illovo.co.za/worldofsugar/internationalSugarStats.htm. [Accessed August 23, 2007.]

ILO (International Labor Organization). 2001. *Promotion of Cooperatives Report 5 (1). Eighty-Ninth Session*. Geneva: ILO.

INCOOP. 2010. "Cantidad de Cooperativas en El Paraguay." http://www.incoop.gov.py/v1/?p=2052.

Indian Sugar Cane Association. 2008. "World Sugar Market Review." http://www.indiansugar.com/briefings/wsm.htm. [Accessed May 10, 2009.]

Instituto de Desenvolviment Agropecuário e Florestal Sustentável do Estado do Amazonas (IDAM). 2015. "Novo Remanso se destaca na produção de Abacaxi." September 1, 2015. http://www.idam.am.gov.br/novo-remanso-se-destaca-na-producao-de-abacaxi. [Accessed December 12, 2016.]

Jesus, Claudio. 2000. *Utopia Cabocla Amazonense*. Canoas (RS): Ulbra.

Jesus, Claudio P., and E. L. Jesus. "Responsabilidade Social e Economia Solidária: Um Modelo de Gestão Auto-Sustentável a Amazônia." *Revista Brasileira de Agroecologia: Resumos do Congresso Brasileiro de Agroecologia* 1(1).

Jesus, Edilza L., and M. Ribeiro. 2008. "Educação e Trabalho em Três Tempos: Distanciamentos e Aproximações." Paper presented to the Seminário de Pesquisa em Educação da Região Sul. Juhno 2008. Itajaí, Santa Catarina.

Johnson, Allen W. 1971. *Sharecroppers of the Sertão: Economics and Dependence on a Brazilian Plantation*. Stanford: Stanford University Press.

Jordan, Carl F., and Charles E. Russell. 1989. "Jari: A Pulp Plantation in the Brazilian Amazon." *GeoJournal* 19(4): 429–35.

Kay, Cristóbal. 2006. "Poverty and Development Strategies in Latin America." *Journal of Agrarian Change* 6(4): 455–508.

———. 2008. "Reflections on Latin American Rural Studies in the Globalization Period: A New Rurality?" *Development and Change* 39(6): 915–43.

Kitaoji, Hironobu. 1971. "The Structure of the Japanese Family." *American Anthropologist* 73(5): 1036–57.

Kroeker, Caroline J. 1996. "The Cooperative Movement in Nicaragua: Empowerment and Accompaniment of Severely Disadvantaged Peasants." *Journal of Social Issues* 52(1): 123–38.

Lafaurie Rivera, José Félix. 2005. *Análisis de Impacto de la Liberación de Precios en el Mercado de la Leche*. Bogotá: Federación Colombiana de Ganaderos. http://portal.fedegan.org.co/pls/portal/docs/PAGE/FNG_PORTLETS/NOTICIASYCOMUNICADOS/DOCUMENTOS/ANALISIS%20DE%20IMPACTO%20LIBERACI%D3N%20PRECIOS%20LECHE%20FINAL%20AG%201%200.PDF. [Accessed June 6, 2012.]

———. 2009. *Sector lácteo colombiano: Una propuesta para reconstruir el sector*. Bogotá: Federación Colombiana de Ganaderos.

Laraia, Roque de Barros, and Roberto da Matta. 1967. *Índios e Castanheiros: A Emprêsa Extrativa e os Índios no Médio Tocantins*. São Paulo: Difusão Européia do Livro.

Lernoux, Penny. 1982. *Cry of the People: The Struggle for Human Rights in Latin America—The Catholic Church in Conflict with U.S. Policy*. New York: Penguin Books.

Levine, Daniel H. 1988. "Assessing the Impacts of Liberation Theology in Latin America." *Review of Politics* 50: 241–63.

Lewis, Paul. 1984. "Paraguay Since 1930." In *The Cambridge History of Latin America*, vol. 8, *1930 to the Present*, ed. Leslie Bethell, 233–68. New York: Cambridge University Press.

Lobo Guerra, Luis. 2001. *Cooperativas de Ahorro y Crédito, Impacto en el Desarrollo Local: Caso: Cooperativa de Ahorro y Crédito Agropecuario COAGRANJA Ltda.* Master of science thesis, Pontificia Universidad Javeriana.

———2002. "Conceptos y Metodología para Evaluar el Impacto de las Organizaciones Cooperativas." In *Gestión y Desarrollo: La Experiencia de las Cooperativas en Colombia*, eds. Juan Manuel Silva and Ricardo Dávila, 101–142. Bogotá: Pontificia Universidad Javeriana, Facultad de Estudios Ambientales y Rurales.

López Cerón, Sonia Carolina, and Enith Ximena Peña Hidalgo. 2005. *El Mercado de Microcrédito Productivo Rural: Comprobación Empírica en el Sur de Santander*. Master's thesis, Pontificia Universidad Javeriana.

MacPherson, Ian. 2002. "Encouraging Associative Intelligence: Co-operatives, Shared Learning, and Responsible Citizenship." Plenary Presentation for the International Association for the Study of Cooperative Education Conference. June 20–23, 2002. Manchester, England.

McGrath, Mary Jean, ed. 1978. *Cooperatives, Small Farmers, and Rural Development*. Madison, WI: University Center for Cooperatives.

MAG-INCOOP. 1997. *Relato Histórico de las Cooperativas en el Paraguay y Base para una Política de Desarrollo Cooperativo*. Asunción: Editora Menno Tec.

Marshall, Robert C., ed. 2010. *Cooperation in Economy and Society*. Lanham: Altamira Press.

Medina Nieto, Felipe. 2002. "La Gestión de la Cooperativa de Villanueva (Santander): Una Organización Solidaria, alrededor de la cual Gira el Desarrollo de un Munícipio." In *Gestión y Desarrollo: La Experiencia de las Cooperativas en Colombia*, eds. Juan Manuel Silva and Ricardo Dávila, 387–415. Bogotá: Pontificia Universidad Javeriana, Facultad de Estudios Ambientales y Rurales.

Mladenatz, Gromoslav. 2003. *História das doutrinas cooperativistas*. Trans. José Carlos Castro, Maira de Graça Leal, and Carlos Potiara Castro. Brasília: Confebras.

Mooney, Patrick H. 2004. "Democratizing Rural Economy: Institutional Friction, Sustainable Struggle, and the Cooperative Movement." *Rural Sociology* 69(1): 76–98.

Mora, Mariana. 2008. "Zapatista Anti-Capitalist Politics and the 'Other Campaign': Learning from the Struggle for Indigenous Rights and Autonomy." In *Latin American Social Movements in the Twenty-First Century: Resistance, Power, and Democracy*, eds. Richard Stahler-Sholk, Harry Vanden, and Glen Kuecker, 151–64. Plymouth: Rowman and Littlefield.

Moran, Hugo Hernan, and José Blas Villalba. 1989. *Estudio Sobre la Experiencia Cooperativa Campesina en Coronel Oviedo.* Asunción: Universidad Católica, Centro Interdisciplinario de Derecho Social y Economía Política.

Murray, Douglas, and Laura T. Raynolds. 2007. "Globalization and Its Antinomies: Negotiating a Fair Trade Movement." In *Fair Trade: The Challenges of Transforming Globalization*, eds. Laura T. Raynolds, Doublas Murray, and John Wilkinson, 3–14. New York: Routledge.

Mutersbaugh, Tad, with D. Klooster, M. C. Renard, and P. Taylor. 2005. "Certifying Rural Spaces: Quality-Certified Products and Rural Governance." *Journal of Rural Studies* 21: 381–88.

Nagel, Beverly Y. 1999. "Unleashing the Fury: The Cultural Discourse of Rural Violence and Land Rights in Paraguay." *Comparative Studies in Society and History* 41: 148–81.

Nash, June, Jorge Dandler, and Nicholas S. Hopkins, eds. 1976. *Popular Participation in Social Change: Cooperatives, Collectives, and Nationalized Industry.* The Hague: Mouton.

Netting, Robert M. 1993. *Smallholders, Householders: Farm Families and the Ecology of Intensive, Sustainable Agriculture.* Stanford: Stanford University Press.

Nimuendajú, Curt. 1920. "Cartas de Belém. Letter to: Ílmo. Sr. Dr. Luiz Bueno Horta Barboso, Diretor do Serviço do Proteção dos Índios. July 23, 1920." In *Etnografia e Indigenismo: Sobre os Kaingang, os Ofaié-Xavante, e os Índios do Pará*, eds. C. Nimuendajú and M.A. Gonçalves, 141–54. São Paulo: Universidad de Campinas.

Nogueira, João Nicédio Alves. 2009. *Cooperativismo como Ferramenta de Desenvolvimento.* Manuscript. Universidade Federal do Ceará.

Noronha, Adolfo Vasconcelos. 1976. *Cooperativismo.* Guarulhos, Brazil: Cúpulo.

Ocampo, José Antonio, and Juan Martín, eds. 2003. *A Decade of Light and Shadow: Latin America and the Caribbean in the 1990s.* Santiago: ECLAC.

OCB (Organização das Cooperativas Brasileiras). 2004. *Cooperativismo Brasileiro: Uma História.* Brasília: OCB.

OCB (Organização das Cooperativas Brasileiras). N.D. "Sescoop." www.brasil cooperativo.coop.br.

Oliveira, Braz de Oliveira. 1979. *Cooperativismo: Guia Prático*. Porto Alegre, Brazil: Fundação para o Desenvolvimento de Recursos Humanos.

Oliveira, Gustavo, and Susanna Hecht. 2016. "Sacred Groves, Sacrifice Zones and Soy Production: Globalization, Intensification and Neo-Nature in South America." *The Journal of Peasant Studies* 43(2): 251–85.

OTISA (oficina Técnica Industrial, S.A.). 2005. "OTISA Sugar Mill." http://oci.nosis.com.ar/ociusers/10434/. [Accessed May 23, 2006]

Page-Reeves, Janet. 1998. "Alpaca Sweater Design and Marketing: Problems and Prospects for Cooperative Knitting Organizations in Bolivia." *Human Organization* 57(1): 83–93.

Panzutti, Ralph. 1999. *Algumas Considerações sobre Cooperativismo e Cooperativas*. São Paulo: International Co-operative Alliance.

Perius, Vergílio Frederico. 2001. *Cooperativismo e Lei*. São Leopoldo: Unisinos.

Piekielek, Jessica. 2010. "Cooperativism and Agroforestry in the Eastern Amazon: The Case of Tomé-Açu." *Latin American Perspectives* 37(6): 12–29.

Pinho, Diva Benevides. 1991. *Grandes Coordenadas da Memória do Cooperativismo Brasileiro*. Vol. II. Brasília: OCB.

———. 2004. *O Cooperativismo no Brasil: Da Vertente Pioneira à Vertente Solidária*. São Paulo: Saraiva.

Porro, Roberto, Robert P. Miller, Marcos R. Tito, Jason A. Donovan, Jorge L. Vivan, Ralph Trancoso, Rudi F. Van Kanten, Jorge E. Grijalva, Bertha L. Ramirez, and Andrés L. Gonçalves. 2012. "Agroforestry in the Amazon Region: A Pathway for Balancing Conservation and Development." In *Agroforestry—The Future of Global Land Use*, eds. P.K. Ramachandran Nair and Dennis Garrity, 391–428. Dordrecht: Springer Netherlands.

Portafolio. 2011. "Piden Prohibir Entrada de Sueros Debido a su Uso para Adulterar la Leche." *Portafolio.co*. January 16, 2011. http://www.portafolio.co/negocios/piden-prohibir-entrada-sueros. [Accessed June 6, 2012.]

Portal Lechero. 2010. "Colombia Dejó de Exportar 500 mil Litros de Leche a Venezuela." *Portalechero.com* August 6, 2010. http://www.portalechero.com/ver_items_descrip.asp?wVarItem=4700. [Accessed August 12, 2011.]

Posey, Darrell A. 2002. *Kayapó Ethnoecology and Culture*, ed. Kristina Plenderleith. London: Routledge.

Quintero Ospina, Juliana Andrea, Jorge Pinilla Vergara, Andrés Felipe Cortes Villa, and Marcela Fajardo Pérez. 2009. *Análisis sobre la Cooperativa Lechera de Antioquia COLANTA*. Trabajo Final de Gerencia Estratégica de Logística. Universidad de Antioquia. Medellín, Colombia.

Ramírez, Luis. 2002. "Fundamentos de Gestión Cooperativa en Procesos de Formulación Estratégica: La Ventaja Cooperativa." In *El Proceso Estratégico en las Organizaciones Cooperativas*. San José, Costa Rica: University of Sherbrooke and Universidad de Costa Rica.

Red Colombiana de Acción Frente al Libre Comercio (RECALCA). 2010. "La Crisis Lechera." July 10, 2010. http://www.recalca.org.co/la-crisis-lechera/. [Accessed June 6, 2012.]

Ricken, José Roberto. 2009. *A Integração Econômica e Social nas cooperativas agropecuárias do Paraná*. Master's thesis, Escola Brasileira de Administração Pública e de Empresas Fundação Getúlio Vargas, Rio de Janeiro.

Rodrigues, Roberto. 2001. *A Segunda Onda Cooperativa: Uma Visão Compartilhada*. São Paulo: Gráfica.

———. 2008. *Cooperativismo: Democracia e Paz "Surfando a Segunda Onda."* São Paulo: OCB.

Rodríguez Alcala, Ramiro, and Thomas Otter. 2008. "How Joint Action Can Increase Production While Contributing to Poverty Reduction: The Case of the Frutika in Paraguay." In *Poverty Reduction That Works: Experience of Scaling Up Development Success*, eds. Paul Steele, Neil Fernando, and Maneka Weddikkara, 147–61. Sterling, VA: Earthscan.

Rodriguez Silvero, R., D. Bernal Bareiro, and C. Patino Cardozo. 2005. *Proceso Regional de Integracion Cooperativa de las Americas (PRICA). Capitulo Paraguay*. Asunción: Conpacoop.

Saes, Maria Sylvia Macchione, Vivian-Lara Silva, Rubens Nunes, and Tamara Maria Gomes. 2014. "Partnerships, Learning, and Adaptation: A Cooperative Founded by Japanese Immigrants in the Amazon Rainforest." *International Journal of Business and Social Science* 5(12): 131–41.

Salazar-Díaz, Delly. 2009. "De la Mano de un Visionario, Es Hoy una Cooperativa Fuera de Serie." *Industria Alimenticia*. September 1, 2009. http://www.industriaalimenticia.com/Articles/Reportaje_Latinoamericano/BNP_GUID_9-5-2006_A_10000000000000653788. [Accessed June 6, 2012.]

Santiesteban Rojas, Diego Fernando. 2008. *Colombia Frente al Posible Tratado de Libre Comercio con Estados Unidos (Análisis por Sectores)*. Bucaramanga: Grupo de Investigación en Desarrollo Empresarial Competitivo–Porter, Universitaria de Investigación y Desarrollo.

Schuh, Edward. 1976. *O Desenvolvimento da Agricultura no Brasil*. Rio de Janeiro: APEC.

Schwartzman, Stephan, and Barbara Zimmerman. 2005. "Conservation Alliances with Indigenous Peoples of the Amazon." *Conservation Biology* 19(3): 721–27.

Sciscioli, Alejandro. 2005. "Paraguay: Sweet Experiment with Organic Sugar." Inter Press Service News Agency. http://www.ipsnews.net/2005/06/paraguay-sweet-experiment-with-organic-sugar/.

SEBRAE. 2009. "Estudo Setorial Pscicultura." http://201.2.114.147/bds/bds.nsf/942E08A72A23F2FB83257627004FE8BD/$File/NT00041FD6.pdf.

Serpa, Horacio. 2011. *Diagnóstico Económico y Social del Departamento de Santander. Plan de Desarrollo Departamental 2008–2011*. Departamento de Santander, Santander, Colombia. http://www.santander.gov.co/santander/documentos/Economia_de_Santander.pdf. [Accessed 14 June 2011.]

Silva, Juan Manuel and Ricardo Dávila, eds. 2002. *Gestión y Desarrollo: La Experiencia de las Cooperativas en Colombia*. Bogotá: Pontificia Universidad Javeriana, Facultad de Estudios Ambientales y Rurales.

Sindicato e Organização das Cooperativas Brasileiras no Estado do Ceará OCB/CE. www.brasilcooperativo.coop.br

Singer, Paul. 2006. "The Recent Rebirth of the Solidarity Economy in Brazil." In *Another Production Is Possible: Beyond the Capitalist Canon*, ed. Boaventura De Sousa Santos, 3–42. London: Verso.

Sorrenson, William J., C. Diarte, and J. Lopez Portillo. 1998. *Economics of No-till Compared to Traditional Cultivation Systems on Small Farms in Paraguay: Policy and Investment Implications*. Asunción: Ministry of Agriculture and Livestock and German Agency for Technical Co-operation.

Souza, Ruan Carlos R., and E. L. Jesus. 2011. "A Identidade e a Memória Como Fatores de Responsabilidade Social do Turismo no Amazonas." Paper presented at the 2nd Congresso de Natureza, Turismo e Sustentabilidade. Cuibá, MT.

Stahler-Sholk, Richard, Harry Vanden, and Glen Kuecker, eds. 2008. *Latin American Social Movements in the Twenty-First Century: Resistance, Power, and Democracy*. Plymouth: Rowman and Littlefield.

Stoesz, Edgar, and Muriel Stackley. 2000. *El Chaco Paraguayo: Tierra de Refugio Patria Adquirida*. Asunción: Macrografica.

Superintendencia de Economía Solidaria de Colombia. 2008. *Circular Básica Contable y Financiera No. 004*. Bogotá: Superintendencia de Economía Solidaria de Colombia, Ministerio de Hacienda y Crédito Público.

Superintendencia Financiera de Colombia. 2011. Retrieved from www.superfinanciera.gov.co. [Accessed June 17, 2011.]

Tafner, Armando Wilson, and Fábio Carlos da Silva. 2010. "Colonização Japonesa, História Econômico e Desenvolvimento Regional de Estado Do Pará." *Novos Cadernos NAEA* 13: 121–52.

TransFair USA. 2009 "Sugar Program." http://www.transfairusa.org/content/sugar_program.php. [Accessed May 5, 2009.]

Última Hora/Dirección General de Estadísticas, Encuestas, y Censos (DGEEC). 2001. *Paraguay 2001—Atlas Estadístico*. Asunción: Última Hora.

United Nations, Department of Economic and Social Affairs, Population Division. 2014. *World Urbanization Prospects: The 2014 Revision, Highlights* (ST/ESA/SER.A/352). New York: United Nations.

USDA (United States Department of Agriculture). 2015. *Paraguay: Gain Report*. Washington, DC: Global Agricultural Information Network, United States Department of Agriculture.

Vanden, Harry E. 2008. "Social Movements, Hegemony, and New Forms of Resistance." In *Latin American Social Movements in the Twenty-First Century: Resistance, Power, and Democracy*, eds. Richard Stahler-Sholk, Harry Vanden, and Glen Kuecker, 21–38. Plymouth: Rowman and Littlefield.

Vásquez-León, Marcela. 1998. "Neoliberalism, Environmentalism, and Scientific Knowledge: Re-Defining Natural Resource Use Rights in Mexico." In *States and Illegal Practices*, ed. Josiah M. Heyman, 233–60. Oxford: Berg.

———. 2009. "One Decade of Drought and Two of Neoliberal Reforms in the Sierra Sonorense: Responses by the Rural Poor." *Journal of Southern Rural Sociology* 24(1): 44–66.

———. 2010. "Free Markets and Fair Trade, Collective Livelihood Struggles, and the Cooperative Model: Two Case Studies from Paraguay." *Latin American Perspectives*. 175(37): 53–73.

Vásquez-León, Marcela, Brian Burke, and Lucero Rádonic. 2009. "Engaging Students in Applied Research: Experiences from Collaborative Research and Learning in Brazil and Paraguay." *Learning and Teaching* 2(2): 46–65.

Vásquez-León, Marcela, and Diana Liverman. 2004. "The Political Ecology of Land-Use Change: The Case of Affluent Ranchers and Destitute Farmers in the Mexican Municipio of Alamos." *Human Organization* 63(1): 21–33.

Weinstein, Barbara. 1983. *The Amazon Rubber Boom, 1850–1920*. Stanford: Stanford University Press.

Weisskoff, Richard. 1992. "The Paraguayan Agro-Export Model of Development." *World Development* 20(10): 1531–540.

World Bank. 2005. *World Development Report 2006: Equity and Development*. Washington, DC: The International Bank for Reconstruction and Development/The World Bank.

———. 2007. *World Development Report 2008: Agriculture for Development*. Washington,

D.C.: The International Bank for Reconstruction and Development/The World Bank.

———. 2015. "Paraguay Country Overview." http://www.worldbank.org/en/country/paraguay/overview. [Accessed June 2015.]

Yamada, Masaaki. 1999. *Japanese Immigrant Agroforesty in the Brazilian Amazon: A Case Study of Sustainable Rural Development in the Tropics*. Doctoral thesis, University of Florida.

Yamada, Masasaki, and Henry L. Gholz. 2002a. "Growth and Yield of Some Indigenous Trees in an Amazonian Agroforestry System: A Rural-History-Based Analysis." *Agroforestry Systems* 55(1): 17–26.

———. 2002b. "An Evaluation of Agroforestry Systems as a Rural Development Option for the Brazilian Amazon." *Agroforestry Systems* 55(2): 81–87.

Yamada, Masasaki, and H. M. L. Osaqui. 2006. "The Role of Homegardens for Agroforestry Development: Lessons from Tomé-Açu, a Japanese-Brazilian Settlement in the Amazon." In *Tropical Homegardens: A Time-Tested Example of Sustainable Agroforestry*, eds. B. M. Kumar and P. K. R. Nair, 299–316. Dordrecht: Springer Netherlands.

Zimmerman, B., C. A. Peres, J. R. Malcolm, and T. Turner. 2001. "Conservation and Development Alliances with the Kayapó of South-Eastern Amazonia, a Tropical Forest Indigenous People." *Environmental Conservation* 28(1): 10–22.

CONTRIBUTORS

Luis Barros holds a B.A. in economics from the University of Brasília and an M.A. in Latin American studies from the University of Arizona, and is currently a Ph.D. candidate in anthropology at the University of Arizona. His master's-level research examined a Brazilian aquaculture cooperative and focused on the development contribution of cooperative action within the socioeconomic and cultural context of the Brazilian northeast. His doctoral research is on the development impacts of participation and grassroots organization on North American indigenous communities. He currently resides in Fortaleza, Ceará, Brazil.

Brian J. Burke is an assistant professor in the Goodnight Family Sustainable Development Department at Appalachian State University. After receiving his Ph.D. in anthropology from the University of Arizona, where he worked extensively with the Bureau of Applied Research in Anthropology (BARA), Brian served as a postdoctoral researcher with the University of Georgia's Coweeta Listening Project and the Coweeta Long Term Ecological Research (LTER) Program. His research aims to support movements for social justice, environmental sustainability, and solidarity economies by examining their ethical visions, strategies, and the challenges they face. Drawing on political economy and political ecology, he is particularly interested in the ways that material and sociocultural dynamics work together to constrain and enable change. Brian's work has included projects on urban environmental activism and appropriate technologies on the U.S.-Mexico border, rural cooperatives in Brazil and Paraguay, alternative economies and ecovillages in Colombia, and local environmental knowledge in southern Appalachia.

Charles C. Cox oversaw technical assistance efforts with agricultural cooperatives as a Peace Corps APCD in Peru and Brazil (1970–1975) after three years as a Peace Corps Volunteer in Ecuador. He earned a BA at Middlebury College in political science/international relations and foreign languages (French, Spanish, Russian) in 1964 and an M.A. at Seattle University in public management in 1976. After graduate school he was an Africa/Middle East/Brazil representative for Volunteer Development Corps (later VOCA, then ACDI/VOCA) CDP projects (1977–1980), then worldwide CDP manager for VDC/VOCA/ACDI/VOCA from 1980 to 2010. He was the founding manager of the worldwide Farmer-to-Farmer Program from 1985 to 2008. That program won a White House award from President George W. Bush in 1989 for successfully helping reduce hunger around the world. He was the founding manager of VOCA's efforts to modernize agriculture, agribusiness, and cooperative operations and results in Central and Eastern Europe and the former Soviet Union (1990 until 2005). He oversaw mission-funded USAID cooperative and agribusiness projects in Panama, Honduras, Colombia, Peru, Bolivia, and Paraguay. He retired from ACDI/VOCA in 2010.

Luis Alberto Cuéllar Gómez has a B.A. in economics and an M.A. in agricultural cooperativsm and rural development from the Centro Internacional de Perfeccionamiento Profesional y Técnico, International Training Center of the ILO in Turin, Italy. He has dedicated nearly 20 years of his 35-year rural agriculture development career to providing technical and management assistance to specialty coffee and cocoa value chain projects, working mainly with cooperatives in Colombia, Brazil, Paraguay, Peru, Central American countries, and Indonesia. He has worked as a project manager for local and international NGOs including ACDI/VOCA. He is currently based in Colombia and works as senior advisor and consultant in specialty coffee and cocoa projects along value chains in Colombia and other countries. His most recent publications on the coffee value chain can be found in Robert Thurson et al. 2013 edited volume "Coffee: A Comprehensive Guide to the Bean, the Beverage, and the Industry." He also has a 2013 International Coffee Organization and World Bank publication, "Case Study G. Mature Cooperative Phase: Growing and Consolidating a Coffee Relationship."

Miguel Ricardo Dávila Ladrón de Guevara is emeritus professor from the Pontificia Universidad Javeriana (PUJ), Colombia. He has an M.A. in rural development (PUJ) and is a member of the PUJ's research group Unit of Solidarity Studies. He is also an associate researcher of the Cooperative Center of Research and Education (CIEC) and the president of the Corporation Red Unicosol. His research focuses on strategy and innovation in solidarity management. His most recent book, *Buenas Prácticas Cooperativas* (Good Cooperative Practices), was published in 2013 by the National University Fund and CIEC, Bogotá D.C. Currently, he is associate researcher in the research project "Conceptualization of the Solidarity Economy in Colombia."

Elisa Echagüe was the director of ACDI/VOCA Paraguay for six years (2004–2009). She is currently the director of the Community Economic Development Sector for the Peace Corps in Paraguay. As an agricultural engineer with a degree from the Universidad Nacional de Asuncion (1990), she has worked with small farmers and smallholder cooperatives throughout Paraguay.

Timothy J. Finan is a professor of anthropology in the School of Anthropology and a research anthropologist at the Bureau of Applied Research in Anthropology at the University of Arizona, where he has 36 years of research, teaching, administration, and outreach experience. His research and outreach activities stretch widely across the globe from Africa to South Asia to Latin America, particularly in Brazil, where he was once a Peace Corps Volunteer. His interests center around society-environment interactions, with specific emphasis on food and livelihood security, climate change adaptation, and governance. His past and current projects include studies of governance and climate change adaptation in the Sahel, resilience to climate change in Bangladesh, and participatory governance in urban Fortaleza, Brazil. He was once a handball player of distinction.

Andrés González Aguilera had the vision and the leadership to build the first sugar mill for organic sugar processing in Paraguay and in the world that markets through fair trade and that is the property of a smallholder cooperative. He is the manager of the Manduvirá cooperative and sugar mill in Arroyos y Esteros, Paraguay, and has considerable experience in cooperative development, fair trade, leadership, and international marketing. He is on the Producer Advisory Council of Sugar PAC, German International Fairtrade, which groups sugar experts from around the world. He is also a founding member of the Federation of Organic and Fair Trade Producers in Paraguay.

Sonia Carolina López Cerón is a professor of finance in the Pontificia Universidad Javeriana in Bogotá. She is currently completing her Ph.D. in development studies at the International Institute of Social Studies of Erasmus University in Rotterdam. She has extensive experience studying cooperatives, with a focus on rural savings and loan cooperatives. Her primary research interests include the social and solidarity economy. She is working from Karl Polanyi's legacy in the understanding of economic integration processes, including nonmonetary integrative mechanisms.

Joana Laura Marinho Nogueira has a degree in law from the University of Fortaleza, a master's degree in international relations from the Pontífica Universidade Católica in Minas Gerais, and a master's degree in cooperative management from Mondragón University in Spain. She has served as international advisor to the Organization of Brazilian Cooperatives (OCB), and since 2012 she has been the coordinator for innovative microenterprise activities in the Ceará Secretariat of Economic Development.

João Nicédio Alves Nogueira has a bachelor's degree in agricultural sciences from the Federal University of Ceará, Brazil, where he specialized in cooperative management. He served as the president of COIGUATU (Cooperativa Agrícola e Industrial de Iguatu), a regional agricultural cooperative, and since 1997 he has been the president of the Ceará State Organization of Brazilian Cooperatives. He currently serves on the board of directors of the Organization of Brazilian Cooperatives (OCB) in Brazil.

Jessica Piekielek is an associate professor of anthropology at Southern Oregon University. She has research and teaching interests in environmental and applied anthropology, including conservation and development in the U.S.-Mexico border region and in Latin America. She received her Ph.D. in anthropology from the University of Arizona in 2009. Her research on small-producer cooperatives in Brazil and Paraguay was published in *Human Organization* (with Brian Burke, 2011, 70:4) and *Latin American Perspectives* (2010, 37:6). Her most recent publication, "Creating a Park, Building a Border: The Establishment of Organ Pipe Cactus National Monument and the Solidification of the U.S.-Mexico Border," appeared in *The Journal of the Southwest* (2016, 58:1).

María Isabel Ramírez Anaya is an agricultural engineer from the Universidad Surcolombiana de Neiva in Huila, Colombia. She has worked in rural development for the last 18 years with organizations including the Centro Nacional de Investigaciones del Café CENICAFE in Colombia, ACDI/VOCA, Coffee Corps, and coffee-producers organizations CAFEFIHUILA and the Grupo Asociativo San Isidro, and as an auditor for Rainforest Alliance certification. For the past nine years she has worked as an auditor for Fair Trade certification (FLOCERT), auditing cooperatives, associations, smallholder businesses, and exporters of coffee, bananas, and golden berries in Colombia and Venezuela. Since 2014 she has also worked with the Italian firm Illycaffé, providing technical support in Colombia and Costa Rica.

Rodrigo Rentería-Valencia is an assistant professor of anthropology at Central Washington University. He holds a Ph.D. in environmental anthropology from the University of Arizona and has published in journals in both Mexico and the United States. He has extensive ethnographic experience among indigenous peoples of the Sonoran Desert and the U.S.-Mexico borderlands, and has conducted research as part of the ACDI/VOCA cooperatives project research team. At the core of his academic contributions is a reconceptualization of hunting as an object of anthropological enquiry in neoliberal times—exemplified by his analysis of the bighorn sheep management program implemented on Tiburón Island, Comcaac territory, Mexico. His current recent interests reside at the intersection of semiotics, political ecology, and the nonhuman turn in the social sciences.

Lilliana Andrea Ruiz Marín is an agricultural engineer from the Universidad Nacional de Colombia—Sede Medellín. She has been working for the past five years at Ecoflora Agro in Colombia. Lilliana has worked on postharvest technologies for fruits and vegetables, and has helped Colombian and Paraguayan farmers' organizations and corporations achieve certification in best agricultural practices, eco-labeling (Ecocert and Ceres), ethical trade (Fair Trade, Global Gap, Ethical Trading Institute), and worker safety (ISO 18000).

Marcela Vásquez-León is the director of the Center for Latin American Studies, an associate professor of anthropology in the School of Anthropology, and an associate research anthropologist at the Bureau of Applied Research in Anthropology at the University of Arizona. Her research and teaching interests include grassroots development and collective organization, environmental and maritime anthropology, rural development, political ecology, and social vulnerability to climate and environmental change. She has conducted research and outreach for over two decades with smallholder agricultural and fishing communities throughout Latin America and the U.S. Southwest on issues related to collective organization, common property resources, and rural development. Most of her work has taken place in northwest Mexico, the U.S.-Mexico border region, Brazil, Paraguay, and Colombia. She has also worked with migrant farm workers and Native American tribes in southern Arizona and with Colombian refugees in the Colombia-Ecuador border. For the past seven years she has been co-director of the Study of the U.S. Institute, which brings mostly indigenous student leaders from a variety of Latin American countries to the United States to learn about, among other things, sustainable development issues and Native American populations in the U.S. Southwest.

INDEX

Page numbers in *italics* indicate illustrations.